Lecture Notes in Social Networks

Series Editors

Reda Alhajj, University of Calgary, Calgary, AB, Canada
Uwe Glässer, Simon Fraser University, Burnaby, BC, Canada

Advisory Board

Charu C. Aggarwal, IBM T.J. Watson Research Center, Hawthorne, NY, USA
Patricia L. Brantingham, Simon Fraser University, Burnaby, BC, Canada
Thilo Gross, University of Bristol, Bristol, UK
Jiawei Han, University of Illinois at Urbana-Champaign, IL, USA
Huan Liu, Arizona State University, Tempe, AZ, USA
Raul Manasevich, University of Chile, Santiago, Chile
Anthony J. Masys, Centre for Security Science, Ottawa, ON, Canada
Carlo Morselli, University of Montreal, QC, Canada
Rafael Wittek, University of Groningen, The Netherlands
Daniel Zeng, The University of Arizona, Tucson, AZ, USA

More information about this series at http://www.springer.com/series/8768

Rokia Missaoui • Sergei O. Kuznetsov
Sergei Obiedkov

Editors

Formal Concept Analysis of Social Networks

 Springer

Editors
Rokia Missaoui
Department of Computer
Science & Engineering
University of Quebec in Outaouais
Quebec, Canada

Sergei O. Kuznetsov
Faculty of Computer Science
National Research University
Higher School of Economics
Moscow, Russia

Sergei Obiedkov
Faculty of Computer Science
National Research University
Higher School of Economics
Moscow, Russia

ISSN 2190-5428   ISSN 2190-5436   (electronic)
Lecture Notes in Social Networks
ISBN 978-3-319-87739-6   ISBN 978-3-319-64167-6   (eBook)
DOI 10.1007/978-3-319-64167-6

Printed on acid-free paper

This Springer imprint is published by Springer Nature
The registered company is Springer International Publishing AG
The registered company address is: Gewerbestrasse 11, 6330 Cham, Switzerland

This book is dedicated to the memory of Prof. Rudolf Wille who was the inventor of Formal Concept Analysis (FCA) and the leader of this community. He was known to be an outstanding person and scientific researcher who influenced the work of many scientists. He passed away on January 22, 2017. We will miss him in our community centered around FCA and its applications.

Foreword

Formal concept analysis emerged around 1980 as a mathematical theory of concepts and concept hierarchies. It was discovered by Rudolf Wille (1937–2017) and his group that a simple and natural mathematical definition leads to abstract objects that are very similar to an old philosophical understanding of "concepts," as expressed in the *Logique de Port-Royal* (l'art de penser, 1662), and that this offered a broad range of new interpretations for modern mathematics. The key ingredient of the *Port-Royal logic* is that concepts are understood as having two parts, an *extension* and an *intension*. The extension of a concept consists of all things that fall under the concept. The intension is the collection of all properties under which the respective concepts fall.

Formal concept analysis should not be misunderstood as a mathematical model of human thought. It is conceivable that formal concept analysis is used as a tool for building such psychological theories, but that would just be another one among many applications. Formal concept analysis does not address a specific problem or aim at a particular application. Its strength is that it offers intuitive access to a mathematical universe and that it is comprehensible because of its similarity to human conceptual thinking. Formal concept analysis was discovered when a group of mathematicians reconsidered their own mathematical research with respect to motivation and meaning, and it caused excitement because it indeed offers new ties between common thought and mathematics.

The distinction between "human" and "formal" concepts is of crucial importance. What is a formal concept, and how it is different from a concept of everyday thinking? A main difference is that formal concepts always are constructed with reference to a concrete, well-restrained set of data. In formal concept analysis one speaks of "objects" and "attributes," but these are always chosen from a well-defined supply. Mathematically speaking, we start each investigation by fixing sets G and M of *objects* and of *attributes* and also define which of these objects *has* which attributes.[1] So what are "objects" and "attributes"? This is arbitrary; it is

---

[1]A more careful presentation would even speak of *formal objects* and *formal attributes*.

not assumed that they are of any specific nature or that they satisfy any particular conditions. All that is required is that objects and attributes are given as "sets" in the mathematical sense and that there is a relation between these sets expressing the object-attribute relationship. No other restrictions are made. It is, for example, not assumed that objects and attributes are different. Nor need they be "atomic" in any sense: often objects and attributes have a complex structure; they may be molecules, languages, algorithms, and so on. When formal concept analysis is used for data analysis, then the data provides the formal context. Sometimes it is necessary to translate given data to this form, and there is a standardized method for doing so, called *conceptual scaling*.

So what is a formal concept? As said before, formal concepts are always understood with reference to three sets (G, M, I), where the elements of G are called *objects*, those of M are the *attributes*, and $I \subseteq G \times M$ is a relation in the mathematical sense, i.e., a set of pairs. A pair (g, m) belongs to the relation I if and only if the object g *has* the attribute m. The three sets together constitute a *formal context* (G, M, I). A *formal concept* of such a formal context (G, M, I) consists of two sets, called the *extent* and the *intent* of the concept. A formal concept of (G, M, I) is a pair (A, B) such that:

- A is a set of objects, expressed as $A \subseteq G$.
- B is a set of attributes, so $B \subseteq M$.
- B consists of precisely those attributes in M that all objects in A share (formally $A' = B$).
- A consists of exactly those objects in G which have all the attributes in B (formally $A = B'$).

There is a natural way of ordering the formal concepts of any given formal context: the *subconcept-superconcept* relation. One formal concept (A_1, B_2) is a subconcept of another one (A_2, B_2), if (and only if) its extent is contained in that of the second, i.e., iff $A_1 \subseteq A_2$. It can easily be argued that this implies that its intent contains the intent of the second, i.e., that $B_2 \subseteq B_1$. This is the mathematical analogue of the philosophical *law of reciprocity* that more general concepts have larger extensions and smaller intensions. But in the mathematical setting, this has a surprising consequence. It can be shown that the ordered set of formal concepts in any case carries an algebraic structure, called a *complete lattice*. This is why the ordered set of all formal concepts is called the *concept lattice* of the underlying formal context (G, M, I).

There is an extensive mathematical theory of complete lattices, and it can be transferred to concept lattices. We obtain, essentially for free, powerful mathematical methods for the treatment of formal concepts. Having such a solid mathematical basis is one of the strengths of formal concept analysis. Moreover, there is a well-tried visualization method for complete lattices, which provides us with expressive and reliable *concept lattice diagrams*, another advantage of the approach. Formal concept analysis also offers powerful algorithms for data analysis. Many variations of the theory have been worked on, allowing, e.g., for the conceptual analysis of relational, fuzzy, or dynamic data. For a mathematical theory, the field is still very young, and certainly only a small anabranch of the great science river. But the number of publications has long exceeded the number of 10,000. So a remarkable supply of results is available. But there are too few systematic presentations focused on specific application domains. This volume on formal concept analysis and social network analysis therefore is very welcome.

As said above, the original motivation for inventing formal concept analysis was not to use it for a particular application but came from the expectation that it could provide meaningful mathematics for a variety of applications. Another important aspect was that it offers a structural and less numeric approach to data. Many of today's data analysis methods rely on numerical values, often expressing some kind of "similarity" by real numbers. But not all types of data are suitable for a numerical description; some require alternative methods. Wille's intention even was that formal concept analysis could contribute to a more human-centered knowledge processing, which still has the virtue of being mathematically rigorous. Numerical methods often result in an "outcome" which compresses the result of an investigation in a few parameters, numbers, or curves. Structural methods, as provided by formal concept analysis, are different. They often unfold the data, making it better accessible to the analyst's judgments and decisions. But these approaches are not exclusive, and formal concept analysis can include numerical considerations whenever appropriate, typically when the amount of data is too large to be unfolded in detail.

This has been the case in the early applications of formal concept analysis to the analysis of networks by Freeman and White for community detection in the 1990s. Later on, Gerd Stumme and his coauthors applied formal concept analysis for detecting trends or shared conceptualizations in so-called folksonomies. They used methods which had been developed for data mining purposes and which would not generate *all* formal concepts of given data but focus on the *frequent* ones. These methods were modified so that they can handle *triadic* formal concepts, which consist of three sets, *extent*, *intent*, and *modus*. They showed that such an approach can indeed reveal characteristic information about social networks. Other authors have contributed their own ideas, and it became necessary to compile the recent trends into this volume. I hope that it will inspire future investigations and applications of formal concept analysis.

Technische Universität Dresden, Germany   Bernhard Ganter

Preface

Introduction

With the wide spread of social networks, the advent of big data, and the increasing number of studies toward the integration of data analysis and social network techniques, it becomes important to get a better insight into existing work, trends, and challenges in pattern discovery from complex social networks in order to efficiently explore and analyze real-life applications.

The book intends to study existing and potential connections between two research areas, social network analysis (SNA) and formal concept analysis (FCA). The papers included in this book show how standard SNA techniques, usually based on graph theory, can be supplemented by FCA methods, which rely on lattice theory.

To the best of our knowledge, there is no book that covers the contents of this volume. However, the two workshops SNAFCA (2007 and 2015) organized by one or all of the coeditors had the same goal. Also, a special issue of the *Social Networks* journal on Social Network and Discrete Structure Analysis published in 1996 covered the use of lattices for the analysis of social data. Linton C. Freeman and Douglas R. White are among the first researchers who investigated the application of FCA for SNA.

This volume is meant to cover the state of the art of the research on the intersection of FCA and SNA in a more systematic and detailed manner than it was done in the workshop proceedings mentioned above. It contains seven chapters written by SNA and FCA researchers. Three chapters are extended versions of selected papers presented at the Social Network Analysis using Formal Concept Analysis Workshop (SNAFCA 2015), which took place in Nerja in August 2015 jointly with the ICFCA 2015 conference. All chapters have been evaluated by two to three reviewers.

As part of the Springer book series *Lecture Notes in Social Networks*, this edited volume, *Formal Concept Analysis of Social Networks*, presents contributions to the following areas: acquisition of terminological knowledge from social networks, knowledge communities, individuality computation, other types of FCA-based analysis of bipartite graphs (two-mode networks), community detection and description in one-mode and multimode networks, multimodal clustering, adaptation of the dual-projection approach to weighted bipartite graphs, and attributed graph analysis.

The goal of the chapter "Knowledge Communities and Socio-Cognitive Taxonomies" by Camille Roth is to show how approaches such as FCA allow the assessment and analysis of actors and their attributes on an equal basis. In the special case of knowledge communities in which the attributes of actors express cognitive properties, one is in presence of joint social and cognitive taxonomies. The chapter also shows that FCA can help solve key typical challenges of community detection in SNA such as group hierarchy and overlapping, temporal evolution and stability of networks.

The chapter "Individuality in Social Networks" by Daniel Borchmann and Tom Hanika defines individuality and introduces a new measure in two-mode (affiliation) networks using FCA by evaluating how many unique groups of users of size k can be uniquely defined by a combination of attributes. The experimental study illustrates the importance of the individuality notion and the additional insights it brings to SNA.

In the chapter "Descriptive Community Detection," Martin Atzmueller presents an overview of recent research about descriptive community and subgroup detection in social networks. Approaches toward the identification of descriptive patterns related to static as well as dynamic relations are described with a focus on attributed graphs. The author also briefly presents his approach to descriptive community detection, which combines subgroup discovery with community detection.

The goal of the chapter "Multimodal Clustering for Community Detection" by Dmitry I. Ignatov, Alexander Semenov, Daria Komissarova, and Dmitry V. Gnatyshak is to present recent progress in object-attribute (OA) biclustering and its extensions to mining multimode communities in social network analysis. Links between clustering coefficients commonly used in SNA community for one-mode and two-mode networks and the OA-bicluster density are also established. Empirical studies with two-, three-, and four-mode large real-life networks show that OA-biclusters are suitable for community detection in multimode cases.

Francesco Kriegel describes a technique for the acquisition of terminological knowledge from social networks in the chapter "Acquisition of Terminological Knowledge from Social Networks." In particular, he provides an extension of the results of Baader and Distel for the deduction of knowledge bases from interpretations in the description logic $\mathcal{MH}$ w.r.t. descriptive semantics and role-depth bounds.

The chapter "Formal Concept Analysis of Attributed Networks" by Henry Soldano, Guillaume Santini, and Dominique Bouthinon studies social and other complex networks as attributed graphs and addresses attribute pattern mining in such graphs through recent developments in FCA. The main idea is to restrict the space of possible pattern extensions in the node set to node subsets satisfying some topological property. To that end, two levels are considered: the abstract and local levels. At the first level, the extension of each pattern is reduced so that the corresponding abstract extension induces a subgraph whose nodes satisfy some connectivity property. At the second level, a pattern has various extensions, each one associated with a connected component of the abstract subgraph attached to the pattern. This leads to abstract closed patterns and local closed patterns as well as abstract and local implications. Interestingness measures for patterns and implications are proposed. Finally, local concepts are linked to network communities, and the illustration of the proposed work is done through the detection and ordering of k-communities in the subgraphs of an attributed network.

In the chapter "A Formal Concept Analysis Look at the Analysis of Affiliation Networks," Francisco J. Valverde-Albacete and Carmen Peléz-Moreno adapt the dual-projection approach of Everett and Borgatti to weighted two-mode networks using an extension of FCA for incidences with values in a special case of semiring. In the case of networks with non-Boolean weights, the dual-projection method is linked to both the singular value decomposition and the eigenvalue problem of matrices with values in such algebras, as in Kleinberg's HITS algorithm. The chapter also introduces extensions of the HITS algorithm to calculate the influence of nodes in a network whose adjacency matrix takes values over dioids, zerosumfree semirings with a natural order. This work shows that the original HITS algorithm is a particular instance of the generic construction and highlights the advantages of working in idempotent semifields, instances of dioids.

The production of this volume would not have been possible without the valuable involvement and efforts of the above contributing authors and the following reviewers: Martin Atzmueller, Jaume Baixeries, Karell Bertet, Mohamed Bouguessa, Dmitry Ignatov, Mehdi Kaytoue, Francesco Kriegel, Léonard Kwuida, Jurgen Lerner, Amedeo Napoli, Henry Soldano, and Francisco J. Valverde-Albacete. We highly appreciate the efforts and commitment of all the authors and reviewers. We would also like to express our gratitude to Christopher T. Coughlin and his team members from Springer USA for their help in the preparation of this volume.

Gatineau, QC, Canada   Rokia Missaoui
Moscow, Russia   Sergei O. Kuznetsov
Moscow, Russia   Sergei Obiedkov
March 2017

Contents

Contributors

Martin Atzmueller Tilburg Center for Cognition and Communication (TiCC), Tilburg University, Tilburg, Netherlands

Research Center for Information System Design (ITeG), University of Kassel, Kassel, Germany

Daniel Borchmann Chair of Automata Theory, Technische Universität Dresden, Dresden, Germany

Dominique Bouthinon Université Paris 13, Villetaneuse, France

Dmitry V. Gnatyshak National Research University Higher School of Economics, Moscow, Russia

Tom Hanika Knowledge & Data Engineering Group, Research Center for Information System Design (ITeG), University of Kassel, Kassel, Germany

Dmitry I. Ignatov National Research University Higher School of Economics, Moscow, Russia

Daria Komissarova National Research University Higher School of Economics, Moscow, Russia

Francesco Kriegel Institute of Theoretical Computer Science, Technische Universität Dresden, Dresden, Germany

Carmen Peláez-Moreno Department of Signal Theory and Communications, Universidad Carlos III de Madrid, Leganés, Spain

Camille Roth Sciences Po, Médialab, Paris, France

Centre Marc Bloch Berlin e.V., Berlin, Germany

Guillaume Santini Université Paris 13, Villetaneuse, France

Alexander Semenov National Research University Higher School of Economics, Moscow, Russia

Mobile TeleSystems PJSC, Moscow, Russia

Henry Soldano Université Paris 13, Villetaneuse, France

Museum National d'Histoire Naturelle, Paris, France

Francisco J. Valverde-Albacete Department of Signal Theory and Communications, Universidad Carlos III de Madrid, Leganés, Spain

Knowledge Communities and Socio-Cognitive Taxonomies

Camille Roth

1 Introduction

A significant portion of the state of the art in social network analysis (SNA) has been typically devoted to the characterization of groups of *similar* actors [68]—oftentimes denoted as *communities*. While similarities and communities can be defined in very diverse ways [6, 23], the literature can roughly be divided into two main classes of approaches, depending on whether groups stem from cohesive interactions or from cohesive affiliations. This dichotomy in turn refers to two distinct types of graphs and relations. On one hand, a large number of methods rely on *interaction networks*, which are purely social networks insofar as nodes are strictly actors and links indicate actor–actor relationships: actors know each other, they work with each other, they talk to each other, etc. Formally, this corresponds to monopartite graphs or the so-called one-mode networks. On the other hand, SNA scholars have made use of *affiliation networks* [68, Chap. 8] where nodes may be of two types: either actors or social attributes of some sort—be it an event, an organization, a team, an issue, an opinion, an interest, etc. Here, relationships denote the affiliation, in the broad sense, of an actor to an attribute; formalisms are based on bipartite graphs, or two-mode networks.

Approaches based on interaction networks traditionally seem to constitute the bulk of the literature on social group characterization from relational data in SNA [28, 68]. They focus on the shape and structure of relationships *between* actors and appear to pay little attention to the cognitive and property-based aspects

C. Roth (✉)
Sciences Po, Médialab, 84 rue de Grenelle, 75007 Paris, France

Centre Marc Bloch Berlin e.V., Friedrichstrasse 191, 10117 Berlin, Germany
e-mail: camille.roth@sciencespo.fr

© Springer International Publishing AG 2017
R. Missaoui et al. (eds.), *Formal Concept Analysis of Social Networks*,
Lecture Notes in Social Networks, DOI 10.1007/978-3-319-64167-6_1

1

of communities as such (even though they may indirectly uncover affiliation communities, i.e., social groups defined by similar attributes, for instance, as a result of homophily, or the fact that similar people interact more with one another [48]). Furthermore, many approaches based on affiliation networks nonetheless fall back on an interaction network by transforming two-mode into one-mode data, i.e., by building an actor network whose links denote shared affiliations among actors. Typical cases include co-appearance in a same movie, co-authorship of a same scientific article, or co-membership in a same team. Some further approaches aim at truly uncovering groups in the two-mode network [31], yet in general an actor-centric view eventually appears to prevail. In other words, the composition of the detected communities and, subsequently, the validity of the results are principally discussed in terms of actors, whereas attributes essentially remain in the background as an instrumental helper: somehow, semantic similarity is used as an indirect tool to uncover implicit interactional patterns.

This is where, we contend, lies one of the most crucial assets of dual approaches such as formal concept analysis (FCA [33]) with respect to SNA: the possibility of describing hybrid communities of actors and attributes in a simultaneous manner, without giving priority on one mode over the other, while tackling several of the key challenges raised by community detection in traditional SNA.

By focusing on knowledge communities and, more precisely, by emphasizing the possibility of formalizing the notion of *epistemic community* (EC), at the interface between SNA and FCA, this chapter aims at showing how FCA may particularly contribute to SNA in uncovering and describing social groups based on cognitive affiliation patterns. To this end, we first recall how structural approaches have formalized the notion of interactional community, discussing in particular the main quantitative issues and qualitative connections with sociological analysis. We then explain how FCA enables, by contrast, the description of social groups which are characterized by attribute similarity and for which it is more straightforward to use affiliation networks. We show how FCA still captures many important community features of interest to SNA. We illustrate this stance with a series of empirical examples.

2   Communities in Interaction Networks

2.1   *Explicit vs. Procedural Methods*

Algebraic Definitions of Social Groups   Formal apprehension of the sociological notion of "community" principally stems from SNA [22], all the more as social interactions have progressively occurred in an increasingly networked fashion [69]. Historically, the introduction of graph theory in sociometry [27] paved the way to the first mathematical analyses of communities: the so-called sociogram of a given group of actors, which describes relationships such as acquaintances, friendships, collaborations, or exchanges, could be represented as a graph. Then, the

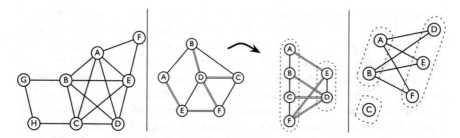

Fig. 1 Explicit, algebraic definitions of groups. *Left:* $\{A, B, C, D, E\}$ and $\{A, E, F\}$ are cliques, $\{B, C, G, H\}$ is a 2-clique. *Middle,* Cartwright and Harary's decomposition [13] (positive or negative links are, respectively, represented by *single* or *double lines*). *Right:* structural equivalence classes of this network include $\{A, B\}$, $\{C\}$, and $\{D, E, F\}$

abstract study of its algebraic and topological structure could reveal the existence of "real" communities, by matching a given qualitative community definition with quantitative graphic properties.

The notion of *clique* [47] has played a foundational and prototypical role in this endeavor. Cliques shall sound familiar to FCA scholars, whose formal concepts are maximal bicliques of a bipartite graph isomorphic to the object-attribute matrix. Interactional cliques are even simpler patterns: they configure subsets of individuals who are *all* connected with one another, i.e., complete subgraphs of the interaction network (see an illustration in Fig. 1). Cliques can thus be seen as the most basic and strongest cohesive community unit. In practice, however, cliques larger than a dozen of actors are relatively rare. Furthermore, as may be the case with formal concepts, their computation, representation, and even interpretation often prove difficult. SNA thus quickly introduced less rigid notions of communities, starting with n-cliques [46] which allow for a looser connectivity among individuals belonging to the same group (they have to be at most at distance n from each other).

Methods which are more global and holistic were also proposed very early to partition a given network into various sub-communities, rather than just exhibiting local patterns such as cliques. Building upon the so-called balance theory introduced in psychology [37] and which allows for either positive or negative relationships between actors (friends/foes, i.e., valued links), Cartwright and Harary [13] were among the first ones to formalize communities at the network-level with their structure theorem. In a nutshell, when some composition laws on relationships hold (namely, foes of friends are foes), they showed that it is possible to split nodes into two groups such that intra-group (resp. inter-group) connections are positively (resp. negatively) valued. Here, communities follow from antagonistic rather than similar interactional configurations. Multiple refinements of this approach have later been introduced [17, 18, 20, 21], focusing, for instance, on the role of triads—again a very local pattern.

Beyond these foundational milestones, SNA has developed over the previous decades a very rich and diverse set of definitions of social groups, where patterns directly match explicit mathematical expressions. Many contributions within this research program are variously evoked in [29, pp. 152–153], [68, Chaps. 7, 9, and 10], [30, pp. 743–744], or [22, pp. 206–207]. Let us mention, for instance, the notion of "equivalence class" of individuals, which describes groups of actors connected to other actors in an equivalent manner. This notably includes *structural equivalence* [45] where actors of the same class share exactly the same neighbors, *regular equivalence* [71] where actors of the same class are linked in a similar way to actors of another class, or *automorphic equivalence* [25], where actors of the same class occupy positions which are exactly interchangeable in the network (their labels may be exchanged without changing the relational structure). In a different direction, the *structural cohesiveness* of a set of actors [70] is defined as the number of individuals which have to be removed in order to get disconnected components, i.e., such that there exists at least one pair of individuals who are not indirectly connected through a chain of links. A group with a structural cohesiveness of k is called a *k-component*: here, communities are groups such that links between actors exhibit some redundancy.

Procedural Methods and Approximate Patterns Formalization does not necessarily imply quantification. In this respect, most of these algebraic approaches were fueled by mathematical sociologists who initially worked on case studies based on small datasets stemming from ethnographic observation, thereby featuring a limited number of actors. As a result, they are essentially adapted to small-sized networks and structures [50] since the number of patterns can grow quickly. How to deal, for instance, with the thousands of cliques which a small network of a hundred of nodes may contain; and what to deduce from their observation?

A more recent stream of research focused in all generality on the quantitative and large-scale study of the topology of social (and non-social) networks. This stance gained momentum during the 2000s, thanks to the joint availability of powerful computational resources and large relational datasets (even if this phenomenon could already be partly perceived as early as the 1970s [2, p. 116]). Under the term "community detection," this literature addresses the issue of the discovery of cohesive structures in large graphs by applying data mining techniques developed to a large extent by computer scientists and statistical physicists [28]. Within this stream, groups or communities are consensually seen as aggregates of actors in the network: "*groups of vertices within which connections are dense, but between which connections are sparser*" [51]. This is aligned with a classical SNA definition: "*its members should have many relations with each other and few with non-members*" [2, p. 121]. Concretely, these approaches are based upon procedural methods and thus tend to blur the distinction between the formal definition of what these "dense" groups are and the algorithm which enables their detection. In contrast with explicit and closed mathematical definitions where "a *group/community* is a set of actors such that [···]," dense group patterns are almost entirely defined by the procedure— all the more when algorithms are stochastic and results vary from an execution to the other. This allows for scalability and, often, compactness of the partitions, to the expense of interpretability.

These algorithms may diversely feature the iterative construction of a series of embedded graph partitions, either by gathering structurally close individuals into

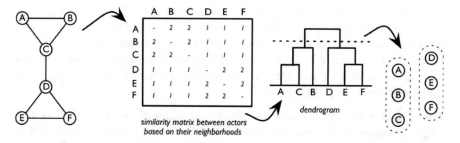

Fig. 2 Partition creation algorithm. *From left to right:* toy network; similarity matrix based on the number of node-independent paths between actors (i.e., paths involving strictly distinct node sets, except for the extremities); iterative construction of the dendrogram based on these similarities (progressively gathering nodes with highest similarities, randomly choosing in case of equality); groups of the partition found at the cut-off stage of the dendrogram, marked by a *dashed line*

increasingly larger groups [14, 72] or by dividing the whole graph into increasingly smaller groups [34]. This procedure is traditionally denoted as *hierarchical clustering* [68], it may be represented as a dendrogram; various criteria such as modularity [52] are then available to decide which partition to choose and where to cut the dendrogram (see Fig. 2 for a toy example). Other procedures can be based on network exploration [7], possibly inspired by percolation processes in order to find community boundaries [54], or holistic methods such as spectral decomposition based on some global properties of the graph adjacency matrix [12, 56].

2.2 Structural Properties of Groups

Structural methods may go beyond the mere partitioning of nodes: they may further be used to describe group structure in itself, i.e., the relationships between groups. Blockmodeling methods, for one, generalize partitioning by reducing the social graph into a meta-graph of groups called *blockmodel*, where nodes represent groups and links describe their relationships.

At the group level, more broadly, we may identify three classical qualitative phenomena which are an important and current research issue in SNA: (1) hierarchies between groups, (2) multiple membership of actors in groups, and (3) temporal dynamics of groups.

Group Hierarchies SNA makes it generally easy to describe social group orders and hierarchies, first and foremost by relying on set inclusion. A group can be "below" or "more specific than" another one if the former is included in the latter: a partial order may be defined where, say, $\{A, B\}$ and $\{B, C\}$ are included in $\{A, B, C\}$ while $\{A, B\}$ and $\{B, C\}$ cannot be compared with one another. Some methods naturally and implicitly define such an order: dendrograms configure increasingly finer partitions, while k-components are included in k'-components when $k' < k$. Traditionally, the resulting hierarchical structure is a *tree* comparable to Aristotelian taxonomies (as in the traditional classification of scientific disciplines: e.g., "scientists" > "biologists" > "molecular biologists" > ...). Hierarchies may also be defined among items of a partition, especially when interactions are directed or valued: [18] uses link asymmetry to define levels between groups, such that *"admiration flows up levels"* as a consequence of differences in the underlying actor prestige or centrality [68, Chap. 5].

Group Overlap Beyond partitions where individuals are meant to belong to a single group (as is the case with equivalence classes), a somewhat small part of the literature has addressed the question of multiple membership [3, 9, 26, 32, 54, for instance]. Here, actors may belong to one or more groups which can in turn partially overlap. While the relevance of taking into account such overlap is sometimes debated (e.g., [29, p. 153]), the relative weakness of scholarly interest in this issue may also be explained by concrete hurdles, such as how to properly justify thresholds triggering multiple membership, or how to deal with the potential combinatorial complexity.

Group Dynamics By definition, interactional analysis of social groups steers clear of intensional properties: in a dynamic perspective, this means that the old sociological question of the perpetuation of social groups[1] is appraised through the stability of interactional structures across time rather than the persistence of their attributes. Typically, inter-temporal correspondence may be assessed longitudinally (groups at t are associated with groups with similar members at t' [19, for instance]) or dynamically (the stability of relationships between t and t' defines the group, as in [49, 53], thereby assuming that social entities only exist by way of their temporal stability [1]). We shall show below how FCA brings a particular added value for this and the above issues, especially in the context of knowledge communities.

3 Reuniting Structure and Content

3.1 Affiliation Networks, Social Circles, and FCA

As mentioned in the introduction, interactional network analysis provides a robust set of methods to define social groups, yet by overlooking a priori their non-structural properties. In this way, since interactional SNA does not rely on intensional properties, it may fail to render the most semantic and cognitive aspects of communities—unless one assumes a strong redundancy between structural and

---

[1]"The most general case in which the persistence of the group presents itself as a problem occurs in the fact that, in spite of the departure and the change of members, the group remains identical. We say that it is the same state, the same association, the same army, which now exists that existed so and so many decades or centuries ago. This, although no single member of the original organization remains." [64, p. 667]

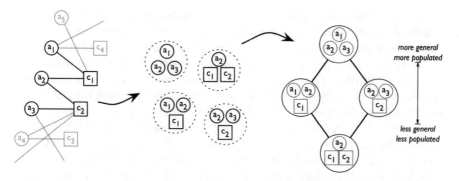

Fig. 3 Bipartite graphs and lattices. *Left:* toy bipartite graph: actors $a_1, a_2, a_3, \dots$ and cognitive properties $c_1, c_2, \dots$. *Middle:* formal concepts corresponding to the black portion of the previous graph. *Right:* lattice-based hierarchical representation

non-structural properties. As such, a social group featuring semantic or cognitive affinity may only be found indirectly if the similarity is manifest in the interactional structure, for example, because of homophily: for instance, scientific collaboration networks exhibit some disciplinary cohesiveness [34, 57]. Semantic labels for Interactional groups are usually labelled a posteriori, if at all, and often by hand. Moreover, larger groups such as schools of thoughts, epistemic communities, interest circles and, more broadly, socio-cognitive groups may not correspond univocally to a single, well-defined interactional community.

The branch of SNA based on affiliation networks appears here as a robust relational framework able to combine structure and semantics. Technically, affiliation networks are bipartite graphs, where actors on one side are distinguished from affiliations on the other side (Fig. 3, left panel). A link may only connect an actor and an affiliation. This formalism is additionally dual, as are social circles [10], in the sense that affiliations are linked to actors just as actors are linked to affiliations.

Social circles are thus explicitly codified in the data: a single affiliation already constitutes an intensional group which denotes the shared participation in an event, membership in an organization, interest for a topic, adhesion to a belief. In this respect, looking for groups in affiliation networks may also be understood as the task of uncovering new (implicit) actor groups from the multiple intersections of social circles, which are thus seen as (explicit) intensional groups [5, 8, 32, 44]. From the viewpoint of SNA, this stance enables both a structural and a cognitive description of communities, which is the cornerstone of describing *socio-cognitive taxonomies*, i.e., joint taxonomies of actors and taxonomies of cognitive attributes— be it in the context of scientists working on research topics, bloggers posting about some issues, activists discussing political matters. Bipartite graphs are isomorphic to binary relations and to labeled hypergraphs (indeed, actor nodes affiliated with the same attribute in a bipartite graph univocally correspond to a labeled hyperedge)— the closeness with FCA is straightforward when considering actors as objects and affiliations as attributes.

While several studies aim specifically at detecting group patterns in bipartite graphs [43], they often tend to consider affiliations as an instrumental rather than fundamental feature. More precisely, many seem to discard the inherent duality either *ex ante*, by focusing on an actor–actor network derived from the original bipartite graph (through a projection of the two-mode network onto one of the modes), or *ex post*, by computing groups of actors with similar properties then discussing the validity of the detected groups principally in terms of actors.

Typically, FCA appears to be one of only a few current methods which aim at maintaining the duality of actors and affiliations along the whole process, from pattern detection to taxonomy interpretation. With respect to the above-mentioned SNA techniques (Sect. 2.1), it also relies on an explicit definition of what a group is, rather than relying on a procedural definition. We will discuss below how FCA also addresses the above-mentioned classical SNA challenges—dealing with group hierarchy, overlap, and dynamics. The resulting computational complexity is also an issue, which has been partly addressed by introducing the first practical application of stability [41] in the very case of socio-cognitive taxonomies and knowledge communities [42, 61].

3.2 Formal Concepts as Epistemic Communities

Before that, we first explain the plain application of FCA on affiliation networks. Formally, we consider the affiliation network as a pair of sets of actors $\mathscr{A}$ and cognitive properties $\mathscr{C}$ (described by, e.g, n-grams, lexical tags, topics, representations, etc.), i.e., agents and notions (or "concepts" in the generic sense of the word), and a binary relationship between them, $\mathscr{R} \subseteq \mathscr{A} \times \mathscr{C}$. The *intent* A' of a set of actors $A \subseteq \mathscr{A}$ is the intersection of all sets of cognitive properties associated with actors of A, i.e., $A' = \{c \in \mathscr{C} | \forall a \in A, a\mathscr{R}c\}$; dually, the *extent* C' of a set of cognitive properties C is the intersection of all actor sets associated with properties of C, i.e., $C' = \{a \in \mathscr{A} | \forall c \in C, a\mathscr{R}c\}$. Applying successively "$'$" yields a closure operator. For all subsets $A \subseteq \mathscr{A}$ and $C \subseteq \mathscr{C}$, (A'', A') and (C', C'') are called *formal concepts* and, equivalently, are maximal bicliques in the bipartite graph of the affiliation network.

In the context of knowledge communities, an efficient qualitative interpretation of formal concepts/biclique patterns consists in considering these socio-semantic groups as *epistemic communities* (EC). Introduced in [63] and later refined by [36] and used by many social scientists afterwards [15, 16], this notion essentially corresponds to actor groups who (1) share some interest for a certain set of topics or beliefs and (2) have a common goal of knowledge creation while obeying to some set of given rules agreed upon in the underlying community. In the very minimal sense, an EC may be formalized as a pair of agents and topics such that all agents share all topics; that is, a biclique in the bipartite affiliation network $(\mathscr{A}, \mathscr{C}, \mathscr{R} \subseteq \mathscr{A} \times \mathscr{C})$. Each EC thus algebraically defined corresponds to a socio-cognitive group which is the closure of a set of actors or equivalently of cognitive properties—a socio-semantic pattern. See illustration in Fig. 3—middle.

Lattices and Socio-Cognitive Taxonomies This formalism addresses several of the issues exposed in Sect. 2.2 regarding interactional groups. In particular, it enables a hierarchical representation of groups through the natural inclusion-based partial order on formal concepts. Conceptually, this hierarchy induces a generalization/specialization relationship: it may be represented as a *lattice* [5]. The most general ECs (largest actor sets/extents, smallest attribute sets/intents) are found towards the top, while the most specific ECs are at the bottom (most specific extents, largest intents). See illustration in Fig. 3—right. This configures a socio-cognitive taxonomy relevant to social epistemology—for one, it is useful to represent distributed cognition activities [38] in a given knowledge production system, in particular the distribution of topics over actors.

Moreover, lattices configure non-Aristotelian taxonomies: ECs partially overlap. Of course, individuals may belong to more than one EC but, more importantly, ECs may also have more than one parent. Arguably, this property makes lattice-based taxonomies closer to cognitive categories, where ECs may simultaneously be subsets of *several* more general ECs.

Finally, it is possible to track the dynamics of these taxonomies by following the evolution of actor sets associated with a given attribute set, thus echoing the ambition of Simmel regarding the persistence of social groups (footnote 1). Note that this approach also inherits a drawback typical of community detection methods based on explicit definitions, especially in the case of cliques: computational complexity. Even for a small number of actors and properties, the number of ECs and the lattice size can be dramatically large [33], easily running in the thousands. This problem is typically critical for SNA scholars, who rarely use cliques, if any. In the next section, we discuss concrete strategies to tackle these issues efficiently.

4 Applications

From the viewpoint of FCA, knowledge communities typically feature either a significant number of actors, or of notions, or both: it is thus key to explain and emphasize how FCA can be of practical use despite combinatorial complexity, especially to compete or keep up with some of the above-mentioned SNA approaches, most notably those based on procedural methods. Data reduction is here a crucial issue, both in terms of input or output, i.e., at the level of the primary data or the computed results.

4.1 Datasets

We present three earlier empirical applications: two are related to scientific communities, one features political activists and motions. These case studies were diversely introduced in [42, 60, 62]: more specific details on each of them may be found in the respective references. In the meantime, FCA has been increasingly applied to groups of actors sharing some properties (e.g., [4, 55]).

In all cases, the empirical material consists of text documents describing, to some extent, who writes about or is interested in what. Actors of the corresponding affiliation networks are identified as document authors, while cognitive attributes are terms extracted from the plain text. A link between actor a and term c occurs whenever a authored a document mentioning c. In this respect, epistemic communities/formal concepts observed in these empirical case studies are strictly speaking socio-lexical patterns.

As the number of individual terms in the original data is always very large, especially with regard to FCA, we systematically apply some filtering relying on simple natural language processing (NLP) techniques. We lemmatize words, exclude stop-words, and eventually focus on the most frequent terms, additionally selecting the most meaningful ones with the help of a domain expert. The number of actors appears to be generally more tractable, yet when it is too large (as in the zebrafish case), we show how simple sampling strategies can be used. See Table 1 for basic statistics regarding the datasets.

The *zebrafish community* case study gathers embryologists who worked on an animal model called "zebrafish" over the years 1990–2003. This period corresponds to the early development of the field, whose population grew approximately tenfold [60]. Data was gathered from the publicly available bibliographical database MEDLINE by querying papers whose abstract includes "zebrafish"—assuming that in most cases authors who work on this animal would necessarily evoke the term in their abstract. The *ECCS* dataset focuses on scholars working on complex systems, focusing on the two first editions of the European Conference on Complex Systems, in 2005 and 2006. The conference organizers kindly provided us with submitted abstracts to both conferences, which we all used in the original study [42]. Finally, the *political motions* example is based on the six roadmaps submitted by six groups of members of the French socialist party towards the internal elections at their Congress in 2008 [62]. In these texts, signatories defend their vision of where

Table 1 Basic figures describing the size of the respective empirical datasets in terms of original documents, unique actors, and NLP-extracted terms

Dataset		Documents	Actors	Terms
ECCS	2005	194	413	92
	2006	187	401	109
Political motions		6	6	85
Zebrafish	1990–1995	533	1094	66
	1998–2003	4080	9689	67

the party should go in the coming years. We consider motions as actors of the corresponding affiliation network, i.e., six nodes; we also keep 85 pre-processed words appearing at least 32 times in the whole corpus.

4.2 Socio-Cognitive Taxonomies

Hierarchy and Overlap We first use the zebrafish case study to illustrate the hierarchy and overlap between groups which is made possible by FCA-based socio-cognitive taxonomies. The period 1990–1995 already features a thousand of actors and 66 attributes—something which yields about eight million ECs and, admittedly, can get neither drawn nor interpreted. A first reduction strategy may consist in operating at the level of the input data by sampling the actor set, assuming that a random portion of the population would still render a faithful taxonomy of the whole community (if needed, removed actors may later be assigned to the computed taxonomy). We use an affiliation subnetwork including a random share of 20% of the population and use it to compute a formal concept lattice made of about 200k ECs. This still represents a sizeable number of ECs, and further reduction may be needed. A second strategy may consist in filtering the output, for instance, by conserving formal concepts according to some relevance criterion. The so-called iceberg lattices [67] have been classically used, whereby a certain portion of the top of the lattice is conserved, assuming that this portion corresponds plausibly to the most interesting or the most meaningful part of the taxonomy. Extent size, i.e., population size of ECs, is a popular criterion; distance to the top may also be used. In Fig. 4, we show such a truncated lattice for the period 1990–1995, together with the last period 1998–2003, to exhibit the temporal evolution.

Let us first focus on the general structure for a given period, say 1998–2003, after the zebrafish community reached some maturity. This picture describes succinctly its main research axes, their representativity, overlaps, and hierarchical relationships. To put it shortly, we see three pillars: (1) comparative studies occupy an important position (*human/mouse/homologous genes*), (2) the study of the nervous system, around the *dorsal* and *ventral* plates, also gathers a certain proportion of scholars, and (3) systemic studies linked to signaling during embryonic development are well-represented (*signal/pathway/growth/receptor*).

Temporality Additionally, we may compare lattices for different periods within the same knowledge community. By focusing on identical attribute groups (intents) across time, FCA makes it possible to render the temporal evolution and relative stability of, at the macro-level, socio-cognitive taxonomies and, at the meso-level, social groups—a key issue in SNA as well. Here, however, the inter-temporal correspondance of groups will be based on attributes rather than interactions.

In practice, we represent evolution by coloring ECs corresponding to a given intent according to the growth of their population share (extent representativity).

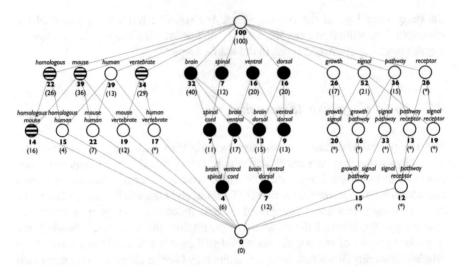

Fig. 4 Truncated lattice of the zebrafish community [59, 60]. *Lines* represent direct hierarchical (inclusion) relationships. Each EC is described by a word list (the intent of the formal concept) and its actor size as a percentage of the underlying population (extent of the formal concept) in *bold* for 1998–2003 and in *parentheses* for 1990–1995. *White disks* indicate growth from the first to the last period, *solid black disks* indicate decline, and *dashed disks* correspond to relative stagnation (arbitrarily defined as variation smaller than ±15%). Percentage sums go over 100% as a result of multiple membership. By definition the *bottom node* gathers all attributes (not listed here)

We see in Fig. 4 that comparative studies have expanded within the zebrafish community, together with the analysis of systemic signals, which echoes a general trend in molecular biology at that time, whereas studies centered around the embryonic nervous system are progressively fading (also a general trend in the surrounding fields). While showing the diversity of the distribution of cognitive tasks within the community, this comparison also demonstrates that it enjoys a remarkable stability, given that the underlying population grew tenfold between the two periods.

Approximation We now turn to the political motion dataset to illustrate reduction strategies further. We use stability [41], a criterion which removes redundancy across the whole lattice and has been widely used in the FCA community since its inception [11, 61]. It indeed constitutes a robust approach to deal with potentially large lattices such as those emerging from empirical social data, while still paying attention to smaller yet plausibly meaningful and representative patterns which would be filtered out by top-down approaches based on, e.g., iceberg-like criteria.

In a nutshell, the (extensional) stability of a given formal concept (A'', A') is formally defined as $\sigma(A'', A') = |\{B \subseteq A'' | B' = A'\}|/2^{|A''|}$, i.e., the proportion of subsets B of the actor set A'' of a given formal concept whose intent B' is identical to A'. Put differently, this criterion measures how much the existence of a given EC depends on its actors. The higher the σ, the more stable the EC, and the more likely it will be presented in the final results.

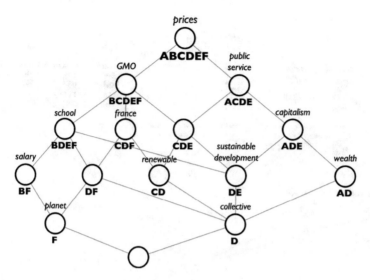

Fig. 5 Stabilized concepts for the political motion dataset (after [62]). Letters *A–F* refer to the six motions. Here, for the sake of readability, cognitive attributes of nodes are labeled in a parsimonious way: we only indicate terms which are *added* to formal concept intents from *top* to *bottom*; for instance, the node "(france, *CDF*)" corresponds to an EC gathering motions *CDF* and terms "france," "GMO," and "prices" (terms are thus inherited from *bottom* to *top*)

Figure 5 presents a reduced taxonomy based on stability. It remains readable while featuring specialized groupings quite deep down from the top. At the most general level, the structure exhibits the omnipresence of issues related to purchase power—all motions talk about "prices"—or GMOs, used by all but motion A. Yet, we also see progressively smaller groupings: for instance, school-related issues (used by motions B, D, E, and F), and then, even lower, joint use of "sustainable development" by D and E, or "salary" by B and F. Dually, we see that motion D is present in almost all ECs by addressing issues present in all other motions.

Combining Both We finally use the ECCS case to illustrate the application of both principles: temporality and approximation (see [42] for more details). Figure 6 shows the 15 most stable concepts for lattices computed over all authors in each year.

On the whole, the main pillars of this scientific field revolve around "networks," "models," and their "dynamics," as well as, to a weaker extent, "structure" and "distribution" (which, in this context, most often refer to scale-free distributions). At the global level, structures for both time periods are relatively comparable. A finer examination reveals some differences: several specific ECs (subconcepts) disappeared in 2006 ({network, dynamics}, {dynamics, model, process}, {dynamics, process}, and {information}) while others appeared ({interaction}, {network, social}, {model, agent}, and {simulation, model}). Focusing on specific intents also provides extra information on the epistemological evolution in 2006: for instance,

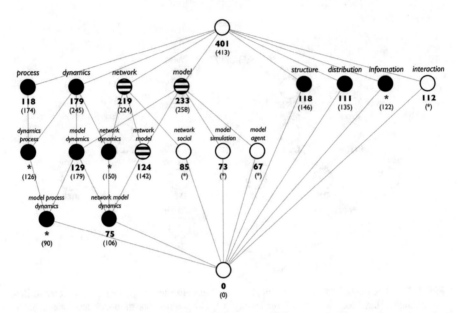

Fig. 6 Stabilized concepts for ECCS-2005 and ECCS-2006 (after [42]). Figures are absolute size of ECs

the EC on {network, dynamics} does not exist anymore on its own, while {network, dynamics, model} still does, suggesting that network dynamics is entirely subsumed by dynamic network models.

5 Concluding Remarks

Beyond the diversity of the SNA literature on the detection of groups, we could draw a fundamental dichotomy between interaction-based and affiliation-based group definitions. In the very case of scientific communities, social scientists argue for a similar dichotomy [40] between *"taxonomic collectives,"* which are relevant at a high level of observation, and interaction groups, in which actors are embedded and which are also relevant at the local level to understand actor behavior. The case studies presented here show how the notion of EC and, behind this, FCA applied to affiliation networks provide a description of the configuration of actor groups in knowledge communities in a manner at least similar to what is possible through classical interactional SNA, while taking actor attributes into account.

With these dichotomies in mind, we can sketch some of the issues where FCA could create a most relevant bridge over SNA for the study of knowledge networks. This includes, first and foremost, the study of the correlation between affiliation and interaction communities. In other words, describe to what extent socio-cognitive communities are also strongly cohesive in interactional terms, how taxonomic collectives may be interaction groups and whether epistemic communities do cover various interaction communities. More generally, do one-mode communities correspond to two-mode communities or formal concepts [58]? Here, some empirical answers have been recently proposed in this direction [65] relying on the so-called alpha concept lattices [66].

Second, on a more practical and theoretical level, the development of approximation strategies is key to guarantee the acceptance of FCA by SNA scholars. This is all the more true in socio-cognitive contexts where result interpretability, both in terms of social groups and in terms of cognitive taxonomies, needs to be manually tractable and therefore involve a sensibly limited number of categories. Stability-based pruning is an option among many, especially in the case of noisy data stemming from social behavior [39]. The design of scalable selection criteria [35] adapted to a socio-cognitive context could be another promising direction of research.

Third, much remains to be done with respect to the dynamics, for instance, by digging further the intensional stability of communities across time. As could be seen here, socio-cognitive taxonomies plausibly evolve slowly, even in the case of a high turnover of actors from a period to the other. On the FCA side, this touches the issue of inter-lattice comparisons [73] and their temporal analysis [24, 74], even though this area remains relatively nascent in FCA. On the side of SNA, group evolution is mainly assessed through a single-network lens. Most likely, appraising simultaneously the joint evolution of social and cognitive patterns, possibly to the point where social groups are even defined by the dynamic stability of socio-cognitive patterns, would constitute a fruitful contribution to social analysis.

Acknowledgements The present contribution partially relies on ideas introduced in a book chapter originally published in French and entitled "Communautés, analyse structurale et réseaux socio-sémantiques" [59].

References

1. Abbott, A.: Things of boundaries. Soc. Res. **62**(4), 857–882 (1995)
2. Alba, R.D.: A graph-theoretic definition of a sociometric clique. J. Math. Sociol. **3**, 113–126 (1973)
3. Arabie, P., Carroll, J.D.: Conceptions of overlap in social structure. In: Freeman, L.C., White, D.R., Romney, A.K. (eds.) Research Methods in Social Network Analysis, pp. 367–392. George Mason University Press, Fairfax, VA (1989)
4. Balamane, A., Missaoui, R., Kwuida, L., Vaillancourt, J.: Descriptive group detection in two-mode data networks using biclustering. In: Proc. of 2016 IEEE/ACM Intl. Conf. on Advances in Social Networks Analysis and Mining (ASONAM). IEEE Computer Society, San Francisco (2016)
5. Barbut, M., Monjardet, B.: Algèbre et Combinatoire, vol. II. Hachette, Paris (1970)
6. Bell, C., Newby, H.: Community Studies: An Introduction to the Sociology of the Local Community. Allen & Unwin, London (1972)

7. Blondel, V.D., Guillaume, J.L., Lambiotte, R., Lefebvre, E.: Fast unfolding of communities in large networks. J. Stat. Mech. Theory Exp. **2008**, P10008 (2008)
8. Boeck, P.D., Rosenberg, S.: Hierarchical classes: model and data analysis. Psychometrika **53**(3), 361–381 (1988)
9. Bonacich, P.: Using boolean algebra to analyze overlapping memberships. Sociol. Methodol. **9**, 101–115 (1978)
10. Breiger, R.L.: The duality of persons and groups. Soc. Forces **53**(2), 181–190 (1974)
11. Buzmakov, A., Kuznetsov, S.O., Napoli, A.: Is concept stability a measure for pattern selection? Proc. Comput. Sci. **31**, 918–927 (2014)
12. Capocci, A., Servedio, V., Caldarelli, G., Colaiori, F.: Detecting communities in large networks. Physica A **352**, 660–676 (2005)
13. Cartwright, D., Harary, F.: Structural balance: a generalization of Heider's theory. Psychol. Rev. **63**, 277–292 (1956)
14. Clauset, A.: Finding local communities in networks. Phys. Rev. E **72**, 026132 (2005)
15. Cohendet, P., Créplet, F., Dupouet, O.: Organisational innovation, communities of practice and epistemic communities: the case of Linux. In: Economics with Heterogeneous Interacting Agents, pp. 303–326. Springer, Berlin (2001)
16. Cowan, R., David, P.A., Foray, D.: The explicit economics of knowledge codification and tacitness. Ind. Corp. Chang. **9**(2), 212–253 (2000)
17. Davis, J.A.: Clustering and structural balance in graphs. Hum. Relat. **20**, 181–187 (1967)
18. Davis, J.A., Leinhardt, S.: The structure of positive interpersonal relations in small groups. In: Berger, J., Zelditch, M., Anderson, B. (eds.) Sociological Theories in Progress. Houghton Mifflin, Boston, MA (1970)
19. Doreian, P.: On the evolution of group and network structure. Soc. Netw. **2**, 235–252 (1979)
20. Doreian, P., Mrvar, A.: A partitioning approach to structural balance. Soc. Netw. **18**(2), 149–168 (1996)
21. Doreian, P., Mrvar, A.: Partitioning signed social networks. Soc. Netw. **31**, 1–11 (2009)
22. Edling, C.R.: Mathematics in sociology. Annu. Rev. Sociol. **28**, 197–220 (2002)
23. Elias, N.: Towards a theory of communities. In: Bell, C., Newby, H. (eds.) The Sociology of Community: A Selection of Readings. Routledge, London (1974)
24. Elzinga, P., Wolff, K., Poelmans, J.: Analyzing chat conversations of pedophiles with temporal relational semantic systems. In: Proc. 1st IEEE European Conference on Intelligence and Security Informatics, pp. 242–249. Odense, Denmark (2012)
25. Everett, M.G.: Role similarity and complexity in social networks. Soc. Netw. **7**, 353–359 (1985)
26. Everett, M.G., Borgatti, S.P.: Analyzing clique overlap. Connections **21**(1), 49–61 (1998)
27. Forsyth, E., Katz, L.: A matrix approach to the analysis of sociometric data: preliminary report. Sociometry **9**(4), 340–347 (1946)
28. Fortunato, S.: Community detection in graphs. Phys. Rep. **486**, 75—174 (2010)
29. Freeman, L.C.: The sociological concept of 'group': an empirical test of two models. Am. J. Sociol. **98**(1), 152–166 (1992)
30. Freeman, L.C.: Un modèle de la structure des interactions dans les groupes. Rev. Fr. Sociol. **36**, 743–757 (1995)
31. Freeman, L.C.: Finding social groups: a meta-analysis of the Southern women data. In: Breiger, R., Carley, K., Pattison, P. (eds.) Dynamic Social Network Modeling and Analysis, pp. 39–97. The National Academies Press, Washington, DC (2003)
32. Freeman, L.C., White, D.R.: Using Galois lattices to represent network data. Sociol. Methodol. **23**, 127–146 (1993)
33. Ganter, B., Wille, R.: Formal Concept Analysis: Mathematical Foundations. Springer, Berlin (1999)
34. Girvan, M., Newman, M.E.J.: Community structure in social and biological networks. PNAS **99**, 7821–7826 (2002)

35. Gnatyshak, D., Ignatov, D.I., Semenov, A., Poelmans, J.: Gaining insight in social networks with biclustering and triclustering. In: Aseeva, N., Babkin, E., Kozyrev, O. (eds.) Perspectives in Business Informatics Research BIR 2012: 11th Intl. Conf., Nizhny Novgorod, Russia, Sept 24–26, pp. 162–171. Springer, Berlin (2012)
36. Haas, P.: Introduction: epistemic communities and international policy coordination. Int. Organ. **46**(1), 1–35 (1992)
37. Heider, F.: Attitudes and cognitive organization. J. Psychol. **21**, 107–112 (1946)
38. Hutchins, E.: Distributed cognition. In: Smelser, N.J., Baltes, P.B. (eds.) International Encyclopedia of the Social and Behavioral Sciences, pp. 2068–2072. Elsevier, Amsterdam (2001)
39. Klimushkin, M., Obiedkov, S., Roth, C.: Approaches to the selection of relevant concepts in the case of noisy data. In: Kwuida, L., Sertkaya, B. (eds.) Proc. 8th Intl. Conf. Formal Concept Analysis. LNCS/LNAI, vol. 5986, pp. 255–266. Springer, Berlin (2010)
40. Knorr-Cetina, K.: Scientific communities or transepistemic arenas of research? A critique of quasi-economic models of science. Soc. Stud. Sci. **12**(1), 101–130 (1982)
41. Kuznetsov, S.: Stability as an estimate of degree of substantiation of hypotheses derived on the basis of operational similarity. Nauchn. Tekh. Inf. **2**(12), 21–29 (1990)
42. Kuznetsov, S., Obiedkov, S., Roth, C.: Reducing the representation complexity of lattice-based taxonomies. In: Priss, U., Polovina, S., Hill, R. (eds.) Conceptual Structures: Knowledge Architectures for Smart Applications: 15th Intl. Conf. on Conceptual Structures, ICCS 2007, Sheffield, UK. LNCS/LNAI, vol. 4604, pp. 241–254. Springer, Berlin (2007)
43. Latapy, M., Magnien, C., Vecchio, N.D.: Basic notions for the analysis of large two-mode networks. Soc. Netw. **30**(1), 31–48 (2008)
44. Lehmann, S., Schwartz, M., Hansen, L.K.: Biclique communities. Phys. Rev. E **78**, 016108 (2008)
45. Lorrain, F., White, H.C.: Structural equivalence of individuals in social networks. J. Math. Sociol. **1**(49–80) (1971)
46. Luce, R.D.: Connectivity and generalized cliques in sociometric group structure. Psychometrika **15**, 169–190 (1950)
47. Luce, R.D., Perry, A.: A method of matrix analysis of group structure. Psychometrika **14**, 95–116 (1949)
48. McPherson, M., Smith-Lovin, L., Cook, J.M.: Birds of a feather: homophily in social networks. Annu. Rev. Sociol. **27**, 415–444 (2001)
49. Mitra, B., Tabourier, L., Roth, C.: Intrinsically dynamic network communities. Comput. Netw. **56**(3), 1041–1053 (2012)
50. Moody, J.: Peer influence groups: identifying dense clusters in large networks. Soc. Netw. **23**, 261–283 (2001)
51. Newman, M.E.J.: Detecting community structure in networks. Eur. Phys. J. B **38**, 321–330 (2004)
52. Newman, M.E.J.: Modularity and community structure in networks. PNAS **103**(23), 8577–8582 (2006)
53. Palla, G., Barabási, A.L., Vicsek, T.: Quantifying social group evolution. Nature **446**, 664–667 (2007)
54. Palla, G., Derényi, I., Farkas, I., Vicsek, T.: Uncovering the overlapping community structure of complex networks in nature and society. Nature **435**, 814–818 (2005)
55. Poelmans, J., Ignatov, D.I., Kuznetsov, S.O., Dedene, G.: Formal concept analysis in knowledge processing: a survey on applications. Expert Syst. Appl. **40**(16), 6538–6560 (2013)
56. Pothen, A., Simon, H.D., Liou, K.P.: Partitioning sparse matrices with eigenvectors of graphs. SIAM J. Matrix Anal. Appl. **11**(3), 430–452 (1990)
57. Rodriguez, M.A., Pepe, A.: On the relationship between the structural and socioacademic communities of a coauthorship network. J. Informet. **2**, 195–201 (2008)

58. Roth, C.: Binding social and semantic networks. In: Proceedings of ECCS 2006, 2nd European Conference on Complex Systems, Oxford (2006)
59. Roth, C.: Communautés, analyse structurale et réseaux socio-sémantiques. In: Sainsaulieu, I., Salzbrunn, M., Amiotte-Suchet, L. (eds.) Faire communautén société – Dynamique des appartenances collectives, pp. 113–128. Presses Universitaires de Rennes, Rennes (2010)
60. Roth, C., Bourgine, P.: Lattice-based dynamic and overlapping taxonomies: the case of epistemic communities. Scientometrics **69**(2), 429–447 (2006)
61. Roth, C., Obiedkov, S., Kourie, D.G.: Towards concise representation for taxonomies of epistemic communities. In: Yahia, S.B., Nguifo, E.M. (eds.) Proc. CLA 4th Intl. Conf. on Concept Lattices and Their Applications. LNCS/LNAI, vol. 4923, pp. 240–255. Springer, Berlin (2006)
62. Roth, C., Cointet, J.P., Obiedkov, S., Romashkin, N.: Analyse textuelle des motions du Congrès de Reims du PS (2008). http://tinyurl.com/39g6lch
63. Ruggie, J.G.: International responses to technology: concepts and trends. Int. Organ. **29**(3), 557–583 (1975)
64. Simmel, G.: The persistence of social groups. Am. J. Sociol. **3**(5), 662 (1898)
65. Soldano, H., Santini, G.: Graph abstraction for closed pattern mining in attributed networks. In: ECAI, pp. 849–854 (2014)
66. Soldano, H., Ventos, V.: Abstract concept lattices. In: Valtchev, P., Jäschke, R. (eds.) Proc. Intl. Conf. on Formal Concept Analysis (ICFCA). LNAI, vol. 6628, pp. 235–250. Springer, Heidelberg (2011)
67. Stumme, G., Taouil, R., Bastide, Y., Pasquier, N., Lakhal, L.: Computing iceberg concept lattices with TITANIC. Data Knowl. Eng. **42**, 189–222 (2002)
68. Wasserman, S., Faust, K.: Social Network Analysis: Methods and Applications. Cambridge University Press, Cambridge (1994)
69. Wellman, B., Carrington, P.J., Hall, A.: Networks as personal communities. In: Wellman, B., Berkowitz, S.D. (eds.) Social Structures: A Network Analysis, pp. 130–184. Cambridge University Press, Cambridge (1988)
70. White, D.R., Harary, F.: The cohesiveness of block in social networks: node connectivity and conditional density. Sociol. Methodol. **31**, 305–359 (2001)
71. White, D.R., Reitz, K.P.: Graph and semigroup homomorphisms on networks of relations. Soc. Netw. **5**, 193–234 (1983)
72. White, H.C., Boorman, S.A., Breiger, R.L.: Social-structure from multiple networks. I: blockmodels of roles and positions. Am. J. Sociol. **81**, 730–780 (1976)
73. Wille, R.: Concept lattices and conceptual knowledge systems. Comput. Math. Appl. **23**, 493 (1992)
74. Wolff, K.: Applications of temporal conceptual semantic systems. In: Wolff, K., Palchunov, D.E., Zagoruiko, N.G. (eds.) Knowledge Processing and Data Analysis. LNAI, vol. 6581, pp. 59–78. Springer, Heidelberg (2011)

Individuality in Social Networks

Daniel Borchmann and Tom Hanika

1 Introduction

Social networks form an integral part of human societies, and their study has been at the core of social science for a long time. It is only recently that mathematical methods have entered the stage, mainly because social networks are now made more explicit than ever due to the availability of social media. This has allowed classical mathematical instruments from graph theory and elsewhere to be applied to social networks—with astonishing results.

One of the first breakthroughs in understanding social networks by means of properties of their graph representations is due to the seminal work by Watts and Strogatz [26]. Here the authors introduce the notion of *small world networks*, encompassing the two simple graph properties of *average shortest path length* and *average local clustering coefficient*. Based on these properties, a graph is said to be a *small world network* if the average shortest path length is small and if the average local clustering coefficient is large. The second seminal result in that direction is the work by Barabási and Albert [3], where social networks are characterized as graphs whose *degree distribution* follows a low-degree *power-law distribution*. It turns out that, surprisingly, both small world networks and power-law distributions describe social networks to a large degree.

D. Borchmann
Chair of Automata Theory, Technische Universität Dresden, Dresden, Germany
e-mail: daniel.borchmann@tu-dresden.de

T. Hanika (✉)
Knowledge & Data Engineering Group, Research Center for Information System Design (ITeG),
University of Kassel, Kassel, Germany
e-mail: tom.hanika@cs.uni-kassel.de

© Springer International Publishing AG 2017
R. Missaoui et al. (eds.), *Formal Concept Analysis of Social Networks*,
Lecture Notes in Social Networks, DOI 10.1007/978-3-319-64167-6_2

In the wake of the results around small world networks, a plenitude of graph-related properties have been reinterpreted as properties of social networks, a popular example being the interpretation of *cliques* in social networks as social groups. However, despite a comparably vast body of research, characterizing all relevant aspects of social networks in terms of mathematical properties of their graph representation has not been achieved to a satisfactory degree. In particular, graphs exist on which existing measures cannot differentiate further, but which intuitively represent qualitatively different social networks.

In this work we want to consider another facet of bipartite social networks, which, as far as we can see, has not been investigated in the literature. This facet is *individuality* in social networks, and by this we intuitively mean the number of unique groups of users a social network has. Note that despite the fact that individuality is concerned with individual users of a network, the measure of individuality we want to investigate in this work is a property of the whole network. It should thus not be confused with notions such as *centrality* or *betweeness*, which apply only to individual vertices instead.

To define the uniqueness of a group of users, we consider the uniqueness of its *milieu* in the given bipartite social network. This intuition of individuality strongly depends on the actual definition of "milieu," a notion that has been discussed in the social sciences before. However, we shall define and employ in this work an interpretation of this word that is different from the one usually used [24].

In a classical representation of social networks as graphs, two users are linked by an edge if and only if they "know" each other in this network. Then the notion of a milieu of a particular user could just be represented as the *neighborhood* of this user in this graph. In this work, however, we want to take up a different stand by representing bipartite social networks as *formal contexts*. These are structures originating from the theory of *formal concept analysis* [9, 27] that allow general investigations of data sets comprising of objects with certain attributes. Using formal contexts, we shall represent a social network as a collection of users with certain properties, where the actual choice of the properties is a matter of modeling. In this way, we can represent various aspects of a social network in a uniform manner.

The main goal of this work is to illustrate that our new notions of individuality are both natural and meaningful. To this end, we shall examine these measures on various real-world data sets, providing evidence that our definitions are reasonable. Even more, we shall show that networks that are similar in terms of their small world character can vary widely when it comes to individuality, suggesting that our new notion expresses properties of social networks that are not covered by the standard notions.

The paper is structured as follows. After revisiting some existing research on mathematical investigations of social networks in Sect. 2, we shall have a closer look on how to represent social networks as formal contexts in Sect. 3. Thereafter, we shall present our notion of *individuality* in Sect. 4, together with the two auxiliary measures of *individuality distribution* and *average milieu size*. An experimental investigation of these new notions follows in Sect. 5. We close with Outlook in Sect. 6.

2   Related Work

Formal concept analysis originated as a subfield of mathematical order theory, more precisely of lattice theory [4]. Lattice theory itself has already been applied to social network analysis, in particular to understanding the clique distribution (among others) in social networks, for example, in [7]. In this work concept lattices were used to analyze the relations between cliques.

Cliques indeed will play a major role in our considerations, and, as already mentioned, cliques have been investigated in the realm of social network analysis before. For example, the clique distribution of social networks was investigated in [28], where the focus was on empirically studying the connection between the power-law distribution of network nodes and the density of cliques. The authors showed to what extent the clique size distribution can be used to estimate the clique density in a social network. In [10] the authors proposed a method to efficiently estimate the distribution of clique sizes from a probability sample of network nodes. However, both works considered uni-modal social networks only. Previous work that also considered clique distributions in bi-modal networks is [22], where it is shown that medium sized cliques are more common in real-world networks than triangles. However, here only cliques in the projected graph were considered, and not in the original bipartite graph.

To the best of our knowledge, individuality in social networks as we consider it in this article has not been studied before as a property of social networks. The only relevant prior work is from the second author [2], on which this article greatly expands.

3   Social Networks as Formal Contexts

Formal concept analysis deals at its core with the representation of complete lattices through *formal contexts*. These are structures $\mathbb{K} = (G, M, I)$ where G and M are sets and $I \subseteq G \times M$ is a binary relation. The standard interpretation of formal contexts is that the set G is a set of *objects*, the set M is a set of *attributes*, and $(g, m) \in I$ signifies that g *has* the attribute m.

Indeed, modeling bipartite social networks as formal contexts is straightforward: consider a social network and identify within this network two sets U and A. We think of the set U as the set of *(interesting) users* of the network and of the set A as the set of *(relevant) attributes* of the users in U. Note, however, that this interpretation of U as a set of users and A as a set of attributes is only one among many possible ones, and there is no restriction on the type of elements contained in these sets.

After having identified the sets U and A, a formal context *representing* a social network is of the form (U, A, I) where $(u, a) \in I$ for $u \in U$, $a \in A$ only if user u *has* attribute a. This representation is also closely linked to considering *bi-modal*

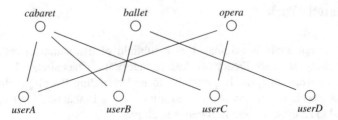

Fig. 1 Small motivational example, called the *music interest social network* (misn)

social networks, i.e., social networks that give rise to a bipartite graph. The benefit of choosing formal contexts over bipartite graphs is that in the former case we can apply methods from formal concept analysis to obtain further insights.

The particular choices of the user set U and the attribute set A are modeling decisions, and finding these sets may not at all be straightforward. For the set U one usually collects all real users of the framework, but other choices—depending on the particular application in mind—are possible. The set A of attributes can contain usual features such as *likes*, *posts*, and *gender*, but can also contain rather "unnatural" features such as other users. In this case, one could define, say, that some user u "has" some other user v as a feature if and only if they are linked in the original social network.

A small example of a social network is given by the bipartite graph in Fig. 1. A formal context representing this network is

$\mathbb{K}_{\text{misn}}$	cabaret	ballet	opera
userA	×		×
userB	×	×	
userC	×		×
userD		×	

In formal contexts we can define two natural *derivation operators* as follows. Let $A \subseteq G$ be a set of objects. Then the set A' of *common attributes* of A is defined as

$$A' := \{m \in M \mid \forall g \in A \colon (g, m) \in I\}.$$

Dually, for a set $B \subseteq M$ of attributes, we define the set B' of *satisfying objects* of B as

$$B' := \{g \in G \mid \forall m \in M \colon (g, m) \in I\}.$$

Note that although both operators are denoted by $\cdot'$, there is usually no danger of confusion, as it is clear from the context whether we are dealing with a set of objects or a set of attributes.

A pair (A, B) is called a *formal concept* of $\mathbb{K}$ if and only if $A' = B$ and $B' = A$. The set A is then called the *extent* and B is called the *intent* of the formal concept (A, B), respectively. Indeed, for each set $A \subseteq G$, the set A is an extent of $\mathbb{K}$ if and only if $A'' = A$. The set of all formal concepts of $\mathbb{K}$ is denoted by $\mathfrak{B}(\mathbb{K})$.

Let us point out the connection of formal concepts to cliques in bipartite graphs: for any formal context $\mathbb{K}$ emerging from a bipartite graph, every formal concept of $\mathbb{K}$ corresponds to a maximal bi-clique in the graph and vice versa.

On the set of all formal concepts $\mathfrak{B}(\mathbb{K})$ we can define a natural order as follows. Let $(A_1, B_1), (A_2, B_2) \in \mathfrak{B}(\mathbb{K})$. Then we say that (A_2, B_2) is *more general* than (A_1, B_1), in symbols $(A_1, B_1) \leq (A_2, B_2)$, if and only if $A_1 \subseteq A_2$. While this definition looks rather asymmetric at first, it turns out that $(A_1, B_1) \leq (A_2, B_2)$ if and only if $B_2 \subseteq B_1$. Moreover, the relation $\leq$ is an *order relation*, and $\mathfrak{B}(\mathbb{K})$ together with $\leq$ forms a complete lattice, the *concept lattice* of $\mathbb{K}$. Conversely, one of the first results of formal concept analysis states that every complete lattice is isomorphic to the concept lattice of some formal context. In this way, formal concept analysis acts as a *representation theory* of complete lattices. Formal concept analysis also allows to link lattice theory to *relational data sets*, as the latter can naturally be represented as formal contexts. In this way, formal concept analysis makes accessible methods from lattice theory for the study of relational data tables.

4   Individuality of Social Networks

We have motivated our notion of individuality by the uniqueness of user milieus. Clearly, this motivation strongly depends on the particular interpretation of the word "milieu", and it is the purpose of this section to provide a formal definition for it. Indeed, modeling a social network by a formal context suggests an immediate definition that is both simple and, as we find, convincing.

Let $\mathbb{K} = (U, A, I)$ be a formal context representing a social network. Then for each user $u \in U$ we define the *milieu* of u simply as the set $\{u\}'$ of attributes common to u. Moreover, if $V \subseteq U$ is a set of users, then the *milieu of V* is the set of attributes common to all users in V, i.e., V'. Using this definition of user milieus, we want to measure the individuality of a social network $\mathbb{K}$ by the amount of milieus that occur in $\mathbb{K}$. Indeed, we shall be a bit more careful here, and propose a notion of *k-group individuality* as a measure to quantify the number of milieus that occur in $\mathbb{K}$ as the milieu of groups of size k, in the sense of how many of the milieus occurring in our social network $\mathbb{K}$ define groups of size exactly k, compared to the number of all groups of size k. Then, the more individuality a social network contains, the more individual groups of a certain size can be defined through their milieu. Conversely, if a social network is quite homogeneous, then defining certain subgroups of individuals by their milieu is improbable.

This approach can naturally be rephrased in terms of formal concept analysis: measuring individuality in $\mathbb{K}$ for user groups of size k is the question of how many subsets $V \subseteq U$ with $|V| = k$ can be expressed in terms of $V = B'$ for some $B \subseteq A$. In other words, we ask for the number of *extents of size k* in $\mathbb{K}$ and use this number to measure the k-group individuality in $\mathbb{K}$. The following definition captures this idea.

Definition 1 Let $\mathbb{K} = (U, A, I)$ be a formal context. Define the set $\text{Ext}_k(\mathbb{K})$ as the set of extents of $\mathbb{K}$ of size k, i.e.,

$$\text{Ext}_k(\mathbb{K}) := \{V \subseteq U \mid V = V'', |V| = k\}.$$

Then the *k-group individuality* $\text{gi}_k(\mathbb{K})$ of $\mathbb{K}$ is

$$\text{gi}_k(\mathbb{K}) := \frac{|\text{Ext}_k(\mathbb{K})|}{\min\{\binom{|U|}{k}, 2^{|A|}\}}. \tag{1}$$

Note that we also normalize by the factor $\min\{\binom{|U|}{k}, 2^{|A|}\}$, because this is the maximal number of k-groups definable by their milieu, and thus allows comparability between individuality of different networks. The used normalization is not optimal, as for k larger than 1 the value of $\text{gi}_k(\mathbb{K})$ rapidly decreases. However, so far the authors are not aware of other normalization approaches.

On a side note, one may also consider the dual measure taking the intents of size k, which would help to measure and describe the individuality of a social network from the attribute point of view.

In terms of measuring the individuality in a social network, the value $\text{gi}_1(\mathbb{K})$ is of particular interest, as this is the percentage of users in this network uniquely determinable by their milieu. In this case, we shall also talk about the *user individuality* $\text{ui}(\mathbb{K}) = \text{gi}_1(\mathbb{K})$ of a social network $\mathbb{K}$.

Using our example from Fig. 1, we first compute the extent sets. As we see in Fig. 2, the concept lattice consists of four elements (apart from the top and bottom ones), and consequently there are four different extents. Indeed we obtain

Fig. 2 Formal concept lattice for $\mathbb{K}_{\text{misn}}$

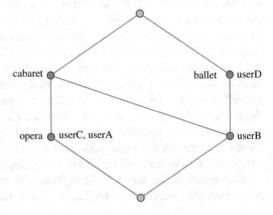

$$\text{Ext}_1 = \{\{\text{userB}\}\},$$

$$\text{Ext}_2 = \{\{\text{userB}, \text{userD}\}, \{\text{userA}, \text{userC}\}\},$$

$$\text{Ext}_3 = \{\{\text{userA}, \text{userB}, \text{userC}\}\}.$$

Therefore, $\text{gi}_1(\mathbb{K}_{\text{misn}}) = \frac{1}{4}$, since only one user has a unique interest that is not covered by another user. We also obtain $\text{gi}_2(\mathbb{K}_{\text{misn}}) = \frac{1}{3}$, demonstrating that in this network the individuality of "pairs" of users is higher than for individual users. Finally, $\text{gi}_3(\mathbb{K}_{\text{misn}}) = \frac{1}{4}$, showing that there is only one group of size three.

The network would be changed considerably if userC would have liked ballet instead of cabaret. In this context, which we want to call misn', there would be three extents of size one and therefore $\text{gi}_1(\mathbb{K}_{\text{misn'}}) = \frac{3}{4}$. Additionally, the number of extents of size two would be four, resulting in $\text{gi}_2(\mathbb{K}_{\text{misn'}}) = \frac{2}{3}$. In short, by not being a copy of the interest of userA, userC can shift the individuality of the network massively by one interest change.

A remark on computing k-group individuality is in order. From the very definition of $\text{gi}_k(\mathbb{K})$, it seems as if computing this value requires to iterate through all subsets of G of size k and check whether they are closed under $\cdot''$. However, using methods from formal concept analysis, the overall effort can be reduced to compute only extents of size *at most* k. More precisely, the algorithm of Next-Closure [9] is able to enumerate closed sets of arbitrary closure operators in a particular order. Exploiting the fact that $\cdot''$ is a closure operator allows us to compute all extents of $\mathbb{K}$ with only polynomial overhead. Furthermore, Next-Closure can be extended to compute only extents of size at most k, further reducing the overall computation costs. A drawback is that Next-Closure cannot be extended to only compute extents of size k, a disadvantage that is not of profound severity, since k-group individuality is usually computed for values $k = 1, 2, \ldots, \ell$ up to some limit $\ell \in \mathbb{N}$.

Note that group individuality also allows detecting the presence of large homogeneous groups, i.e., groups of users with the same milieu. Clearly, such a group of size k exists if and only if $\text{gi}_k(\mathbb{K}) > 0$. In other words, the set

$$\text{gid}(\mathbb{K}) := \{k \in \mathbb{N} \mid \text{gi}_k(\mathbb{K}) > 0\}$$

can be seen as a quantity for the *individuality distribution* in the social network represented by $\mathbb{K}$.

Finally, another aspect of group individuality that we want to consider in this work is the question of how much *information* is necessary to define the milieu of a group of size k. In terms of our modeling of social networks as formal contexts, we reformulate the question to ask how many attributes are necessary *on average* to define a unique group of size k that is itself identifiable through its unique milieu. This gives rise to the following definition.

Definition 2 Let $\mathbb{K}$ be a formal context and let $k \in \text{gid}(\mathbb{K})$. Define the *k-group average milieu size* $\text{ams}_k(\mathbb{K})$ of $\mathbb{K}$ as

$$\text{ams}_k(\mathbb{K}) := \frac{1}{|\text{Ext}_k(\mathbb{K})|} \cdot \sum_{V \in \text{Ext}_k(\mathbb{K})} |V'|$$

For $k \notin \text{gid}(\mathbb{K})$ the value of $\text{ams}_k(\mathbb{K})$ is not defined. It may be set to 0 in those cases if this permits further calculations.

Average milieu size can be naturally linked to *robustness* of group individuality: to deprive a group of k users of being definable in terms of their milieu, on average $\text{ams}_k(\mathbb{K})$ attributes have to be removed from the social network. Consequently, if there are more than $\text{ams}_k(\mathbb{K})$ attributes removed from the network, substantial changes in the k-group individuality should be expected. Verifying this intuition is not within the scope of this work, and is left for future work.

5 Experimental Results

To illustrate our definitions of measuring individuality in social networks, we shall investigate seven different real-world social networks, introduced in Sect. 5.1. We shall see in Sect. 5.2 that all these social networks are indeed small world networks. In Sects. 5.3, 5.4, and 5.5, we examine group individuality, group individuality distribution, and average milieu size of these networks. Finally, we discuss our findings in Sect. 5.6.

5.1 Data and Modeling

In the following we provide short descriptions of the used data sets. The graph properties of all mentioned graphs are summarized in Table 1.

Table 1 Investigated (bi-)partite graphs and their properties

Graph	Vertices in U	Vertices in V	Edges	Edge-density
G_{CM}	40	25	95	0.095
G_{FB}	899	522	7089	0.015
G_{ALNM}	111	134	480	0.032
G_{PLNM}	607	209	5361	0.042
G_{APLNM}	79	188	903	0.061
G_{NB}	1495	367	1746	0.003
G_{SW}	18	14	89	0.35

5.1.1   Club Membership Network (CM) [14]

This data set consists of a bipartite graph describing the affiliations of a set of corporate executive officers to a set of social organizations. This graph consists of 65 vertices representing 40 persons (U_{CM}) and 25 organizations (V_{CM}), as well as 95 edges connecting them. In the following we shall denote this graph by $G_{CM} = (U_{CM} \cup V_{CM}, E_{CM})$.

5.1.2   Facebook-Like Forum Network (FB) [17]

This data set was created by using data from an online community of students from the University of California, Irvine. By using a forum and posting messages to various topics, the students and the topics constitute a bipartite social network. This network consists of a set of 899 users (U_{FB}) and a set of 522 topics (V_{FB}) as well as 7089 edges relating a topic to a user. We shall refer to the resulting graph as $G_{FB} = (U_{FB} \cup V_{FB}, E_{FB})$.

5.1.3   Lange Nacht der Musik (LNM) [20]

This data set stems from an annual cultural event organized in the city of Munich in 2013, the so-called *Lange Nacht der Musik* (Long Night of Music). The corresponding network consists of two bipartite graphs and their intersection. All three of them make use of the same set of vertices, consisting of 1159 users (U_{LNM}) and 212 distinct performances (V_{LNM}).

The first graph records for some users their attendance to performances. We refer to this *attendance graph* by $G_{ALNM} = (V_{ALNM} \cup U_{ALNM}, E_{ALNM})$, where $V_{ALNM} \subseteq V_{LNM}$ and $V_{ALNM} \subseteq V_{LNM}$.

The second graph represents the preferences of some users for where to go during the event. We call this graph the *preference graph* and refer to it in the following as $G_{PLNM} = (V_{PLNM} \cup U_{PLNM}, E_{PLNM})$, where $V_{PLNM} \subseteq V_{LNM}$, and $V_{PLNM} \subseteq V_{LNM}$.

Finally, by intersecting the vertex sets of G_{ALNM} and G_{PLNM} and restricting E_{ALNM} accordingly, we obtain a new graph G_{APLNM} that is the graph of performance attendances where the preferences of the users were known beforehand.

5.1.4   Norwegian Board Members (NB) [21]

This data set was compiled to investigate interlocking directorates among 384 public limited companies in Norway. This network consists of 367 companies (V_{NB}), the set of their 1495 directors (U_{NB}), 1746 edges connecting them (E_{NB}). We shall refer to this bipartite graph by $G_{NB} = (U_{NB} \cup V_{NB}, E_{NB})$.

5.1.5   Southern Women (SW) [24]

A systematic collection observing the social activities of 18 individual women (U_{SW}) over a 9-month period. In this time they attended 14 events (V_{SW}). We shall refer to this graph data set by $G_{SW} = (U_{SW}, V_{SW}, E_{SW})$.

5.2   Small World Network Properties

Graphs arising from social networks empirically satisfy the *small world network property* (SWP), i.e., they expose specific characteristics in terms of local clustering and global separation [5, 25, 26]. With exception of the LNM and NB networks, it is well known that all the networks mentioned in the previous section satisfy SWP to a certain extent. It is the purpose of this section to remind the reader of what those specific characteristics are and what particular values they exhibit on the corresponding networks.

In dealing with networks based on bipartite graphs, so-called *bi-modal networks*, it is common to employ *projections* to obtain the so-called *uni-model* social networks that allow arbitrary links between vertices. While this approach may result in unforeseeable difficulties [29, 30], we shall nevertheless employ it in our work. The main reason for this is comparability: the methods from [26] only apply to uni-modal networks, and projections were used to turn bi-modal networks into uni-modal ones.

Given a bipartite network $G = (U \cup V, E)$, we obtain the *projection* $G^U = (U, E^U)$ of G by the following rule: whenever two users $u_1, u_2 \in U$ share a common neighbor in G, i.e., $\{u_1, v\}, \{u_2, v\} \in E$ for some $v \in V$, then an edge in the projected network G^U will connect them, i.e., $\{u_1, u_2\} \in E$. Then G^U is an undirected graph that corresponds to a uni-modal social network.

Since many observations of network properties are inherited from the network's *degree distribution* [13], it is common to validate the SWP of given networks against a so-called *null model*: to confidently claim that a graph indeed represents a small world social network, the values for local clustering and social separation in the null model should not be larger than in the original network. Here a *null model* for a uni-modal projection of the bipartite social network is represented by a graph that possesses an identical vertex degree distribution but otherwise consists of random connections between the vertices only. To obtain such a null model, we employ the algorithm from [11], which shuffles the edges of the original projection of the bipartite social network while preserving the degree of every vertex. In order to obtain a valid null model, i.e., independent from the edges of the input graph, we shuffle for at least 100 times the number of edges in the input graph [15].

In the following we shall explain in detail how global separation and local clustering are measured by means of *average shortest path length* and *average local clustering coefficients*.

5.2.1 Average Shortest Path

A *path* from u to w in a graph is a sequence of $n \in \mathbb{N}$ vertices successively connected by edges. The *length* of such a path is n. A *shortest path* between nodes u and v is a path of minimal length that starts at u and ends at v.

A social network possessing the small world property must exhibit an *average shortest path length* (ASP) that is low compared to the size of the network. For example, the follower graph of twitter has an average path length of about 4.17 [16], the internet router network has a value of 9.51 [23], and the southern women data set has a value of 1.09 [8].

The results we obtained in our experiment are listed in Table 2. All mentioned bipartite networks exhibit a low average shortest path length in their projected graphs. The numbers vary from 2.01 for the attendance network of LNM to 1.09 in the Southern Women data set. Moreover, in almost all cases the corresponding null model features about the same average shortest path length, as expected for small world social networks, with the only exception being the Norwegian Board Membership graph. For this network the value increases by about 15%. The exceptionality of NB among all data sets will prevail in the later measures.

5.2.2 Average Local Clustering Coefficient

Intuitively, a social network possesses a high local clustering, i.e., users that are connected to a particular user are also likely to be connected themselves. Local clustering in networks is measured by introducing a particular quantity called the *average local clustering coefficient* (ALCC) [26], and every social network must have a comparably high value for this parameter.

The average local clustering coefficient for a graph G can be calculated using the local clustering coefficients C_i for every v_i by $\mathrm{alcc}(G) = \frac{1}{n} \cdot \sum_{i=1}^{n} C_i$, where

$$C_i = \frac{2 \cdot |\{\{v_j, v_k\} \in E \mid v_j, v_k \in N_i\}|}{|N_i| \cdot (|N_i| - 1)}$$

and $N_i := \{v \in V \mid \{v_i, v\} \in E\}$ is the *neighborhood* of v_i in G.

Table 2 Average shortest path lengths (ASP) and average local clustering coefficients (ALCC), alongside the values in a corresponding null model (NM)

Graph	# Edges	Density	ASP	ASP:NM	ALCC	ALCC:NM
G_{CM}	259	0.86	1.14	1.14	0.93	0.92
G_{FB}	123,231	0.30	1.70	1.70	0.69	0.62
G_{ALNM}	1145	0.19	2.01	1.93	0.52	0.31
G_{PLNM}	78,415	0.42	1.63	1.63	0.74	0.70
G_{APLNM}	586	0.20	1.58	1.58	0.71	0.64
G_{NB}	421	0.01	1.34	1.55	0.20	0.01
G_{SW}	138	0.90	1.09	1.09	0.94	0.93

To get a feeling of what certain values of ALCC actually mean for social networks, let us look at some examples: the aforementioned internet router network has an ALCC of 0.03, see [26]. Hence, it would not be considered as a small world social network. In comparison, the twitter followers network has an ALCC of 0.3 [16], which is bigger, but yet not high. Thus, twitter is a social network in which the small world property is not pronounced that much. A good example for a social network with a strong small world property is the one formed by actors using their common movies, which has an ALCC of 0.79, see [26].

Table 2 shows the values of ALCC of the projections of our data sets and of a corresponding null model. Here we observe values between 0.20 for NB and 0.94 for SW, and the values in the null model are lower than in the original networks.

5.2.3 Summary

The investigated data sets clearly exhibit small world network character, with exception of the Norwegian Board Member network, because of its low average local clustering coefficient. Nonetheless, this is a social network, since it is derived from real social data, showing that the heuristic of small world networks has its limits when it comes to identifying social networks. Because of this, it will be even more interesting to see the results for our new individuality measures on this network.

A drawback of our approach to identify small world networks is the usage of projections to obtain uni-modal networks from bi-modal ones. Indeed, in the literature bi-modal social networks are rarely analyzed without transforming them into uni-modal networks, since there are only few methods that can be directly applied to the former. With our new individuality measures we therefore hope to provide a reliable new measure that can be directly applied to bi-modal networks.

5.3 Group Individuality

We present in Table 3 and Figs. 3 and 4 the values of k-group individuality for our data sets for $k = 1, 2, 3, 4$. The largest value of 1-group individuality can be found for the NB data set with 0.96. This was not expected due to the very low value of

Table 3 Experimental results for gi_k for $k = 1, 2, 3, 4$

Graph	gi_1	gi_2	gi_3	gi_4
G_{CM}	0.64	0.04	0.01	0.00
G_{FB}	0.70	0.02	0.00	0.00
G_{ALNM}	0.91	0.02	0.00	0.00
G_{PLNM}	0.69	0.02	0.00	0.00
G_{APLNM}	0.81	0.10	0.00	0.00
G_{NB}	0.96	0.00	0.00	0.00
G_{SW}	0.39	0.08	0.02	0.00

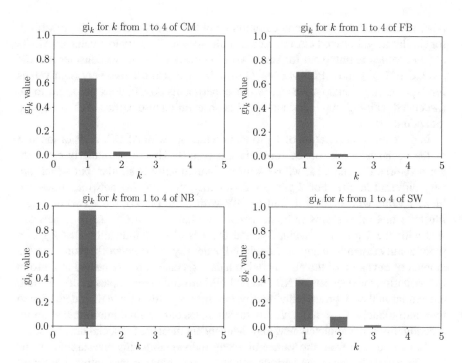

Fig. 3 Group individuality (gi_k) for CM, FB, NB, and SW data sets (from *top left* to *right bottom*)

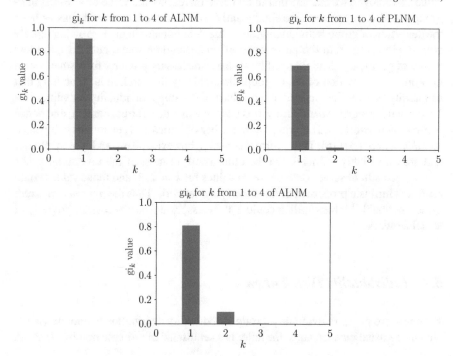

Fig. 4 Group individuality (gi_k) for CM, FB, NB, and SW data sets (from *top left* to *right bottom*)

ASP, which would imply many common neighbors and therefore high probability for similar neighborhoods. A first guess could account the very low value of ALCC for this, which is untypical for small world networks. Yet, if we consider ALNM (ALCC of 0.52) and APLNM (ALCC of 0.71), we also observe very high values for 1-group individuality. Hence, in our experiments ALCC does not seem to be associated with 1-group individuality. This observation also carries over to 2-group individuality.

In general, no correlation of 1-group individuality with ALCC, ASP, or the size of the social network can be found in our results. This is particularly clear for the networks FB and CM, whose k-group individuality is similar, but which are very different in size. For 2-group individuality, the APLNM network shows the highest value with 0.10, followed by SW with 0.08. Indeed, these two data sets illustrate that there seems to be no connection between ALCC, ASP, or network size with the k-group individuality, and there is also no indication that k-group individuality depends in any way on the edge density of the network. Moreover, the amount of deviation of the null model to a data set cannot be connected to k-group individuality: the data sets of NB and APLNM are counterexamples to this, as both are similar in their k-group individuality, but differ significantly in their deviation to their null models. To sum up, all this substantiates our original intuition that group individuality is a completely new and independent measure for social networks.

As can be seen from the values of group individuality, this measure allows us to differentiate between the various networks by exhibiting qualitatively different values. Moreover, we can see that in all cases increasing the value of k results in k-group individuality to decrease significantly. This is indeed expected behavior from the definition of group individuality, since the denominator in gi_k is growing rapidly with k. However, from the perspective of understanding social networks, the low values of gi_k for $k \geq 1$ might itself be seen as a necessary property for a small world network: the formation of large groups definable by their milieu is something that can hardly be expected. Indeed, large values of k-group individuality for values $k > 1$ are usually a sign for *artificiality*: it is easy to generate a formal context, and hence a bi-modal network, with k-group individuality of 1 for $k > 1$, examples being *fixed row density contexts* [6]. Those formal contexts, however, possess a lot of symmetry and are thus highly artificial. On the other hand, in most of the investigated data sets we can still observe some non-zero values for $k = 2, 3$, and those values could represent intrinsic properties of the underlying network. Thus the *presence* of larger groups definable by their milieu could also be associated as a necessary property of social networks.

5.4 Individuality Distribution

In contrast to group individuality, group individuality distribution cannot be visualized in the usual manner, since the latter is a set instead of a simple number. Instead, for every network represented by a formal context $\mathbb{K}$, we computed $gi_k(\mathbb{K})$ for every k from 1 up to the number of users G in the data set. We then identify the value $k_{\max} < |G|$ such that $k_{\max} \in gid(\mathbb{K})$, i.e.,

$$k_{\max} = \max(gid(\mathbb{K}) \setminus \{|G|\}).$$

To visualize $gid(\mathbb{K})$, we then plot its *indicator function* $\mathbf{1}_{\{i \in \mathbb{N} | i < k_{\max}\}}(gid(\mathbb{K}))$. The results are shown in Fig. 5.

The first thing we can read off from the diagrams is of course the corresponding values of $k_{\max}$, the size of the biggest individual group. Furthermore, the density of lines in the plot signifies the existence of groups of various sizes in the network: the more lines are present, the more groups of different sizes exist that are definable through their milieu. From this perspective of its individuality distribution, we perceive PLNM as special, compared to the other networks, because its individuality distribution appears to be very dense. This is also the case for the SW network, because with fourteen users the value for $k_{\max}$ of twelve is also very large. Moreover, comparing PLNM with a data set of comparable size like FB, the structural difference between these networks can be spotted easily: for PLNM the parameter $k_{\max}$ is double as large as for FB. Therefore, even though both networks have similar values for ASP, ALCC, and even for user individuality, the PLNM network seems more interesting than the FB network with respect to their individuality distributions. Indeed, we consider networks with a large value of $k_{\max}$ to be more *interesting* from this point of view.

A more thorough examination reveals that none of the networks exhibits large gaps in their individuality distribution. This is a bit surprising, because one may have expected the existence of very large individual groups and also big gaps in the individuality distribution to the smaller groups, but this is not the case. In general, except for SW, no network exhibits big individual groups definable through their milieu compared to the number of its users.

Finally let us point out that although APLNM is a sub-network of ALNM, their individuality distributions are very similar, although their sizes differ significantly. Based on this observation, one could conjecture that a large part of the individuality of ALNM is already contained in APLNM, or put differently, that most of the individuality of ALNM comes from the APLNM sub-network. However, this conjecture requires further study that is not within the scope of this work.

5.5 Average Millieu Size

The results for our experiments on average milieu size are presented in Table 4 and Figs. 6 and 7. For every data set we computed ams_k for $k = 1, \ldots, 7$. Indeed, for comparing these results with the ones in Sect. 5.3, a maximal value of $k = 4$ would have been sufficient. Yet we observed an interesting peak for the CM data

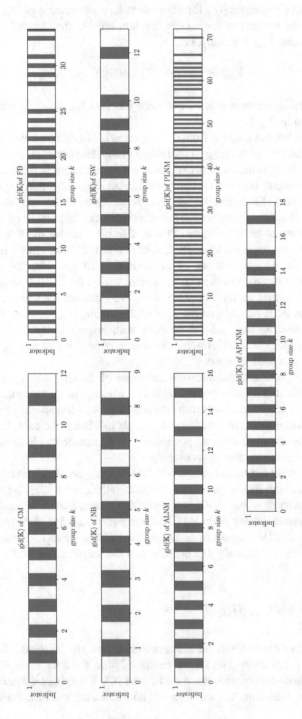

Fig. 5 Individuality distribution of all considered data sets

Table 4 Values of ams$_k$ for $k = 1, \ldots, 7$

Graph	ams$_1$	ams$_2$	ams$_3$	ams$_4$	ams$_5$	ams$_6$	ams$_7$
G_{CM}	4.25	3.50	2.50	1.75	1.00	3.00	2.00
G_{FB}	9.80	3.33	2.55	2.18	1.98	1.69	1.59
G_{ALNM}	4.58	2.17	1.45	1.47	1.25	1.33	1.14
G_{PLNM}	11.5	5.82	4.69	4.07	3.62	3.33	3.13
G_{APLNM}	13.7	5.04	3.39	2.65	2.21	1.91	1.69
G_{NB}	4.97	1.31	1.12	1.00	1.00	1.00	1.0
G_{SW}	7.14	5.23	3.85	2.70	2.43	2.00	2.0

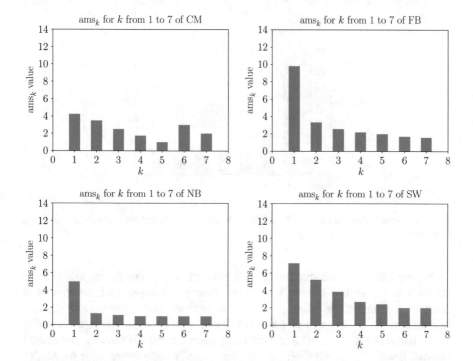

Fig. 6 Average milieu size (ams$_k$) for CM, FB, NB, and SW data sets (from *top left* to *right bottom*)

set, so we decided to show the result as seen. Additionally, we bounded for all plots the maximal value of their y-axis to fourteen to make comparison between the data sets easier.

Comparing the properties of the previous sections, as shown in Tables 1 and 2, to the values of average milieu size in our data sets, again no immediate correlation is visible, suggesting the independence of the introduced measure. In particular, a high value of group individuality does not imply anything on the average milieu size and vice versa. Moreover, the plot for the CM network reveals that, surprisingly, average milieu size does not necessarily need to be monotone in k, as suggested by the other plots.

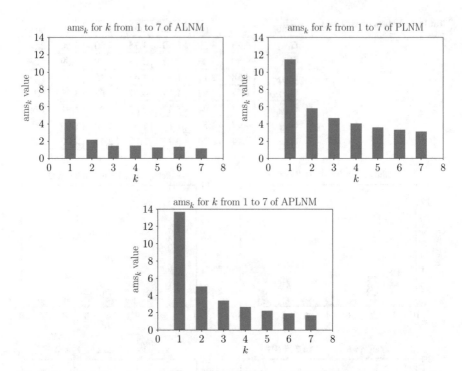

Fig. 7 Average milieu size (ams$_k$) for ALNM, PLNM, and APLNM data sets (from *top left* to *right bottom*)

Among the plots, the one for the NB network stands out for its low average milieu size for groups of size $k > 1$: on average all k-groups have about one attribute in common. Therefore, groups of two or more users rarely have an attribute in common. It is important to point out that in particular the average clustering coefficient is not able to represent this fact: compared to the NB network, the ALNM network has similar average milieu sizes, but a significantly larger value for ALCC.

An interesting observation in the plots is the difference between average milieu size for groups of size 1 compared to larger groups: there is usually a steep decline from the value of ams$_1$ to the one of ams$_2$, say. One may consider a ratio between these values as a measure of how different the milieus of users are compared to those of larger groups of users.

Finally, as explained in Sect. 4, average milieu size can be perceived as a measure for the robustness of the number of k-groups. For this we observe that the APLNM data set reached a value of 14 for ams$_1$, and hence the milieus of milieu-definable users consist on average of 14 attributes. We consequently conjecture the robustness of the user individuality in the APLNM network to be very high, but leave an experimental validation of this hypothesis for future work.

5.6 Discussion and Interpretation

The measures introduced in this work clearly represent facets of individuality in social networks, and it was the purpose of this Sect. 5 to demonstrate the usability and benefit of these quantities. To this end, we discussed various cases where the classical notion of small world networks suggests that two social networks were very similar, but where group individuality and its distribution revealed great structural differences.

The authors can only conjecture the reasons that lead to the observed results. For example, the very high user individuality in the NB network may be explained by strict rules for appointing board members. Especially the very low average milieu sizes for $k > 1$ lead to the impression that there are certain policies in place preventing "clubs" across boards.

The LNM data sets are somewhat special, since they are all intertwined. For example, each of them shows a high user individuality. Using k-group individuality, one may deduce that single users that were tracked during the event were in general more individual in their actions than the ones planning their evening. For the APLNM network, where both attendance and preferences were known, we observe values of user individuality between the ones of ALNM and PLNM. Yet, 2-group individuality is significantly larger in APLNM than in ALNM and PLNM. An interpretation could be that people who planned the evening beforehand are more likely to spend the evening in pairs of two.

To summarize, we claim that the benefit from having an instrument like group individuality is apparent. Furthermore, we assert that there is no method, known to the authors, to get comparable information from a social network.

We want to close this section with a note on the practicability of our approach. We refrained from giving concrete running times for our experiments, mostly because our implementations of the proposed algorithms are preliminary. Showing these values may nevertheless be worthwhile, in particular for arguing that our approach can be applied in practice. Because of this, we show the running times of all our experiments in Table 5. As can be seen from these numbers, the computation times for our data sets never posed a serious problem for the feasibility of our

Table 5 Running times of individual experiments, all times in seconds

	gi	gid	ams
G_{CM}	10.06	9.94	9.84
G_{FB}	170.67	212.94	210.18
G_{ALNM}	13.21	14.04	13.00
G_{PLNM}	1330.58	2302.80	2327.90
G_{APLNM}	13.64	16.00	15.70
G_{NM}	138.56	134.51	140.36
G_{SW}	10.10	10.69	10.44

approach. Moreover, all these running times can be greatly improved by using optimized implementations specifically designed for the fast computation of all formal concepts [1, 18].

6 Conclusions and Outlook

It was the purpose of this work to introduce a new measure on social networks that incorporates the notion of individuality in social networks, an approach that has not been examined before. For this we made use of ideas from formal concept analysis to provide a notion of milieu definability. Based on this, we developed in a natural way the notions of *group individuality*, *individuality distribution*, and *average milieu size*. Conducting experiments on real-world data sets, we were able to show that these new measures were both independent of previously known metrics like ASP and ALCC and allowed differentiating further otherwise similar networks. To sum up, we claim to have shown that the measures of individuality introduced in this work are both natural and meaningful.

This work has only started the study of our individuality measures, and it has not reached its end. For example, so far we have investigated individuality only on real-world networks, where this notion has a natural interpretation. However, we have not even started to look at individuality in networks that do not stem from real-world networks, and we do not know what values of individuality to expect there. In a similar vein, one could ask in how far group individuality is suitable to distinguish real-world networks from artificial ones.

Another aspect that requires further research is the scaling factor for k-group individuality. To improve comparability, we divide the number of extents of size k by $\binom{|G|}{k}$, the theoretical maximal number of such extents. Due to this scaling, k-group individuality is always between zero and one. However, this maximum is never achieved in practice and results in almost-zero values of k-group individuality for larger values of k, making those values virtually useless. Finding a better approach to scale k-group individuality is subject to further investigations.

In our experiments, the running times of our algorithms never posed a problem. However, for larger networks, measuring group individuality can represent a serious challenge: our methods require in the worst case the computation of the whole concept lattice of the representing formal context, and this lattice can be exponentially large. This somehow limits the usefulness of our approach, and further investigations are necessary to explore the possibilities of measuring group individuality of real-world networks.

The networks we have considered in this paper were bi-modal networks from the start, and the actual modeling of finding a suitable attribute set was not an issue. However, for uni-modal networks, finding a suitable set of attributes for a contextual representation may be difficult. To what extent group individuality can be adapted to this kind of networks remains an open problem and is subject to future research.

To establish the small world character of our used data sets, we employ the approach of using null models—something we have not yet done for our individuality measures. One of the main reasons for this is that generating null models for bi-modal networks has received attention from the research community only recently [19], and a proper evaluation is still missing.

A particular kind of social network that is not covered with our contextual representation are the so-called *tripartite* networks, sometimes also called *folk-sonomies* [12]. The corresponding structure in formal concept analysis is the one of a *triadic formal context*, and generalizing group individuality to those structures is also a promising line for future research.

Acknowledgements We thank R. Schaller and B. Ludwig for providing the LNM data set, which is a great addition to this work. Daniel Borchmann gratefully acknowledges support by the Cluster of Excellence "Center for Advancing Electronics Dresden" (cfAED). The computations presented in this paper were conducted by conexp-clj, a general-purpose software for formal concept analysis (https://github.com/exot/conexp-clj). Finally, we would like to thank the reviewers for their insightful comments on this work.

References

1. Andrews, S.: A 'best-of-breed' approach for designing a fast algorithm for computing fixpoints of Galois connections. Inf. Sci. **295**, 633–649 (2015). doi:10.1016/j.ins.2014.10.011
2. Atzmueller, M., Hanika, T., Stumme, G., Schaller, R., Ludwig, B.: Social event network analysis: structure, preferences, and reality. In: Proceedings of IEEE/ACM ASONAM. IEEE Press, Boston, MA (2016)
3. Barabási, A.-L., Albert, R.: Emergence of scaling in random networks. Science **286**(5439), 509–512 (1999)
4. Birkhoff, G.: Lattice Theory. Colloquium Publications, vol. 25, 3rd edn. American Mathematical Society, New York (1967)
5. Boccaletti, S., Latora, V., Moreno, Y., Chavez, M., Hwang, D.-U.: Complex networks: structure and dynamics. Phys. Rep. **424**(4–5), 175–308 (2006). ISSN: 0370-1573
6. Borchmann, D., Hanika, T.: Some experimental results on randomly generating formal contexts. In: Huchard, M., Kuznetsov, S. (eds.) CLA. CEUR Workshop Proceedings, vol. 1624, pp. 57–69. CEUR-WS.org (2016)
7. Freeman, L.C.: Cliques, Galois lattices, and the structure of human social groups. Soc. Netw. **18**(3), 173–187 (1996). ISSN: 0378-8733
8. Freeman, L.C.: Finding social groups: a meta-analysis of the southern women data. In: Dynamic Social Network Modeling and Analysis: Workshop Summary and Papers, pp. 39–97. National Research Council, The National Academies, Washington, DC (2002)
9. Ganter, B., Wille, R.: Formal Concept Analysis: Mathematical Foundations. Springer, Berlin/Heidelberg (1999)
10. Gjoka, M., Smith, E., Butts, C.: Estimating clique composition and size distributions from sampled network data. In: 2014 IEEE Conference on Computer Communications Workshops (INFOCOM WKSHPS), pp. 837–842. IEEE, Toronto (2014)
11. Gkantsidist, C., Mihail, M., Zegura, E.: The Markov chain simulation method for generating connected power law random graphs. In: Proceedings of the 5th Workshop on Algorithm Engineering and Experiments, vol. 111, p. 16. SIAM, Philadelphia (2003)
12. Jäschke, R., Hotho, A., Schmitz, C., Ganter, B., Stumme, G.: Discovering shared conceptualizations in folksonomies. J. Web Semant. **6**(1), 38–53 (2008)

13. Kolaczyk, E.D.: Statistical Analysis of Network Data: Methods and Models. Springer Series in Statistics, p. 386. Springer, New York (2009)
14. KONECT (2016) Club membership network dataset
15. Maslov, S., Sneppen, K.: Specificity and stability in topology of protein networks. Science **296**(5569), 910 (2002)
16. Myers, S.A., Sharma, A., Gupta, P., Lin, J.: Information network or social network?: The structure of the Twitter follow graph. In: Proc. WWW (Companion), pp. 493–498. ACM, Seoul (2014). ISBN: 978-1-4503-2745-9
17. Opsahl, T., Panzarasa, P.: Clustering in weighted networks. Soc. Netw. **31**(2), 155–163 (2009)
18. Outrata, J., Vychodil, V.: Fast algorithm for computing fixpoints of Galois connections induced by object-attribute relational data. Inf. Sci. **185**(1), 114–127 (2012). doi:10.1016/j.ins.2011.09.023
19. Saracco, F., Di Clemente, R., Gabrielli, A., Squartini, T.: Randomizing bipartite networks: the case of the World Trade Web. Sci. Rep. **5**, 10595 (2015)
20. Schaller, R., Harvey, M., Elsweiler, D.: Detecting event visits in urban areas via smartphone GPS data. In: Advances in Information Retrieval. Proc. ECIR. Springer, Cham (2014)
21. Seierstad, C., Opsahl, T.: For the few not the many? The effects of affirmative action on presence, prominence, and social capital of women directors in Norway. Scand. J. Manag. **27**(1), 44–54 (2011)
22. Slater, N., Itzschack, R., Louzoun, Y.: Mid size cliques are more common in real world networks than triangles. Netw. Sci. **2**(3), 387–402 (2014). doi:10. 1017/nws.2014.22
23. Vázquez, A., Pastor-Satorras, R., Vespignani, A.: Internet topology at the router and autonomous system level. In: CoRR (2002). cond-mat/0206084
24. Wasserman, S., Faust, K.: Social Network Analysis. Methods and Applications. Structural Analysis in the Social Sciences. Cambridge University Press, New York (1994)
25. Watts, D.J.: Networks, dynamics, and the small-world phenomenon. Am. J. Sociol. **105**, 493–527 (1999)
26. Watts, D.J., Strogatz, S.H.: Collective dynamics of 'small-world' networks. Nature **393**(6684), 440–442 (1998). ISSN: 0028-0836. doi:10.1038/30918
27. Wille, R.: Restructuring lattice theory: an approach based on hierarchies of concepts. In: Rival, I. (ed.) Ordered Sets: Proceedings of the NATO Advanced Study Institute, pp. 445–470, Banff, 28 August–12 September 1981. Springer, Dordrecht (1982). ISBN: 978-94-009-7798-3
28. Xiao, W.-K., et al.: Empirical study on clique-degree distribution of networks. Phys. Rev. E **76**(3), 037102 (2007)
29. Zweig, K.A.: How to forget the second side of the story: a new method for the one-mode projection of bipartite graphs. In: International Conference on Advances in Social Networks Analysis and Mining, ASONAM 2010, pp. 200–207, Odense, 9–11 August 2010. doi:10.1109/ASONAM.2010.24
30. Zweig, K.A., Kaufmann, M.: A systematic approach to the one-mode projection of bipartite graphs. Soc. Netw. Anal. Min. **1**(3), 187–218 (2011). doi:10.1007/s13278-011-0021-0

Descriptive Community Detection

Martin Atzmueller

1 Introduction

Subgroup discovery [6, 35, 67] aims at identifying interesting descriptive subgroups contained in a dataset—from a compositional network analysis view, aiming at a description given, e.g., by a set of attribute values. The subgroups are identified in such a way that they are interesting with respect to a certain target property. In the context of ubiquitous data and social media, interesting target concepts are given, e.g., by binary variables for obtaining characteristic descriptions of certain phenomena, densely connected graph structures (communities), or exceptional spatio-semantic distributions [4, 8]. This directly bridges the gap to community detection methods [26, 53, 69] that focus on structural aspects of a network/graph, for finding densely connected subgroups of nodes.

This paper, an extended and significantly revised version of [5] presents an organized picture of recent research in subgroup discovery and community detection specifically focusing on attributed graphs. We start with the introduction of necessary background concepts in Sect. 2. After that, we provide a compact overview on prominent methods for community detection, and discuss the exceptional model mining approach. Next, Sect. 3 describes recent work on mining attributed graphs for description-oriented approaches. Then, Sect. 4 summarizes the COMODO algorithm combining both community detection and subgroup discovery in a

M. Atzmueller (✉)
Tilburg Center for Cognition and Communication (TiCC), Tilburg University, Tilburg, Netherlands

Research Center for Information System Design (ITeG), University of Kassel, Kassel, Germany
e-mail: m.atzmuller@uvt.nl

© Springer International Publishing AG 2017
R. Missaoui et al. (eds.), *Formal Concept Analysis of Social Networks*,
Lecture Notes in Social Networks, DOI 10.1007/978-3-319-64167-6_3

description-oriented approach [12, 21], for which we also describe an extension for sequential pattern mining. Finally, we conclude with a summary and point out interesting future directions in Sect. 5.

2 Subgroup Discovery

In general, subgroup discovery can be applied for any standard dataset in tabular form in a straight-forward manner using available efficient algorithms, e.g., [6], as implemented in the VIKAMINE [10, 13] platform. Also, for compositional analysis of social networks, i.e., where nodes have attached attribute information, we can directly apply subgroup discovery for identifying interesting subgroups of nodes according to a given quality measure. The description space is then given by all the compositional variables and their respective value domains. As we will see below, it is also possible to combine a structural with a compositional analysis of a network, resulting in description-oriented community detection using subgroup discovery.

2.1 Patterns and Subgroups

Basic concepts used in subgroup discovery [6, 35, 67] are patterns and subgroups. Intuitively, a *pattern* describes a *subgroup*, i.e., the subgroup consists of instances that are covered by the respective pattern. It is easy to see that a pattern describes a fixed set of instances (subgroup), while a subgroup can also be described by different patterns, covering the subgroup' instances. Below, we define these concepts more formally.

A *database* $D = (I, A)$ is given by a set of individuals I and a set of attributes A. A *selector* or *basic pattern* $\mathrm{sel}_{a_i=v_j}$ is a Boolean function $I \rightarrow \{0, 1\}$ that is true if the value of attribute $a_i \in A$ is equal to v_j for the respective individual. For a numeric attribute a_num whose range is divided into intervals $e_j = [\min_j, \max_j]$ selectors $\mathrm{sel}_{a_\mathrm{num} \in [\min_j; \max_j]}$ can be defined for each interval $[\min_j; \max_j]$ in the domain of a_num. The Boolean function is then set to true if the value of attribute a_num is within the respective interval. The set of all basic patterns is denoted by S.

Definition 1 A *subgroup description* or (complex) *pattern sd* is given by a set of basic patterns sd $= \{\mathrm{sel}_1, \ldots, \mathrm{sel}_l\}$, where $\mathrm{sel}_i \in S$, which is interpreted as a conjunction, i.e., $\mathrm{sd}(I) = \mathrm{sel}_1 \wedge \ldots \wedge \mathrm{sel}_l$, with length(sd) $= l$.

Without loss of generality, we focus on a conjunctive pattern language using nominal attribute-value pairs as defined above in this paper; internal disjunctions can also be generated by appropriate attribute-value construction methods, if necessary [14]. We call a pattern p a *superpattern* (or *refinement*) of a *subpattern* p_s, iff $p_s \subset p$.

Definition 2 A *subgroup (extension)*

$$sg_{sd} := ext(sd) := \{i \in I | sd(i) = \text{true}\}$$

is the set of all individuals which are covered by the pattern *sd*.

As search space for subgroup discovery the set of all possible patterns 2^S is used, that is, all combinations of the basic patterns contained in *S*. Then, appropriate efficient algorithms, e.g., [6] can be applied.

2.2 Interestingness of a Pattern

A large number of quality functions have been proposed in the literature, see [29] for a comprehensive list, in order to estimate the interestingness of a pattern selected according to the analysis task.

Definition 3 A *quality function* $q: 2^S \to \mathbb{R}$ maps every pattern in the search space to a real number that reflects the interestingness of a pattern (or the extension of the pattern, respectively).

Many quality functions for a single target concept (e.g., binary [6, 35] or numerical [6, 43]) trade off the size $n = |ext(sd)|$ of a subgroup for the deviation $t_{sd} - t_0$, where t_{sd} is the average value of a given target concept in the subgroup identified by the pattern *sd* and t_0 the average value of the target concept in the general population. In the binary case, the averages relate to the *share* of the target concept. Thus, typical quality functions are of the form

$$q_a(sd) = n^a \cdot (t_{sd} - t_0), \ a \in [0; 1]. \tag{1}$$

For binary target concepts, this includes, for example, the *weighted relative accuracy* for the size parameter $a = 1$ or a simplified binomial function, for $a = 0.5$. *Multi-target concepts*, e.g., [6, 20, 36, 37] that define a target concept captured by a set of variables can be defined similarly, e.g., by extending a univariate statistical test to the multivariate case, e.g., [20]: Then, the multivariate distributions of a subgroup and the general population are compared in order to identify interesting patterns.

While a quality function provides a *ranking* of the discovered subgroup patterns, often also a statistical assessment of the patterns is useful in data exploration. Quality functions that directly apply a statistical test, for example, the Chi-square quality function, e.g., [6] provide a *p*-value for simple interpretation. However, the Chi-square quality function estimates deviations in two directions. An alternative, which can also be directly mapped to a *p*-value is given by the *adjusted residual* quality function q_r, since the values of q_r follow a large standard normal distribution [3]:

$$q_r = n(t_{sd} - t_0) \cdot \frac{1}{\sqrt{nt_0(1 - t_0)(1 - \frac{n}{N})}} \tag{2}$$

The result of top-k subgroup discovery is the set of the k patterns $sd_1, \ldots, sd_k$, where $sd_i \in 2^S$, with the highest interestingness according to the applied quality function. A subgroup discovery task can now be specified by the five-tuple: (D, c, S, q, k), where c indicates the target concept; the search space 2^S is defined by the set of basic patterns S.

For several quality functions *optimistic estimates* [6, 31] can be applied for determining upper quality bounds: Consider the search for the k best subgroups: If it can be proven that no subset of the currently investigated hypothesis is interesting enough to be included in the result set of k subgroups, then we can skip the evaluation of any subsets of this hypothesis, but can still guarantee the optimality of the result. More formally, an optimistic estimate $oe(q)$ of a quality function q is a function such that $p \subseteq p' \rightarrow (oe(q))(p) \geq q(p')$, i.e., such that no refinement p' of the pattern p can exceed the quality obtained by $(oe(q))(p)$.

2.3   Community Detection

Communities and cohesive subgroups have been extensively studied in social sciences, e.g., using social network analysis methods [66]. Community detection methods can be classified according to several dimensions, e.g., disjoint vs. overlapping communities. Here, actors in a network can only belong to exactly one community, or to multiple communities at the same time. Furthermore, we distinguish between methods that work on extended (attributed) graphs, i.e., including descriptive information about the nodes. Below, we provide an overview on representative methods, including several basic methods working on simple graphs. After that, we elaborate on methods for detecting overlapping communities, before we focus on descriptive methods.

2.3.1   Basics of Community Detection

Wasserman and Faust [66] discuss social network analysis in depth and provide an overview on the analysis of subgroups/communities in graphs, including clique-based, degree-based, and matrix-perturbation-based methods. Furthermore, several algorithms for community detection have been proposed, formalizing the notions of interesting community structures, and introducing the modularity quality measure [51–53]. Fortunato [26] presents a thorough survey on the state-of-the-art community detection algorithms in graphs, focusing on detecting *disjoint* communities.

For assessing the quality of a community, usually not only the density of the community is assessed but also the connection density of the community is compared to the density of the rest of the network [53]. For the modularity measure the number of connections within the community is compared to the statistically "expected" number based on all available connections in the network. Besides modularity, prominent examples of community quality measures include, for example, the segregation index [27] and the inverted average out-degree fraction [70].

2.3.2 Detecting Overlapping Communities

Overlapping communities allow an extended modeling of actor–actor relations in social networks: Nodes of a corresponding graph can then participate in multiple communities. This is also typically observed in real-world networks regarding different complementary facets of social interactions [55]. A general overview on algorithms for overlapping community detection is provided by Xie et al. [69]. For example, clique percolation methods proposed in [55, 56] detect k-cliques and then merge them into overlapping communities. Xie and Szymanski [68] present methods that extend the idea of label propagation [58]. Lancichinetti et al. [40] describe an approach for overlapping and hierarchical community structure using a local community metric. The presented metric itself is computed locally but still assesses a global clustering. Further statistical and local optimization algorithms include the COPRA [30] algorithm by Gregory using label propagation of neighboring nodes until a consensus is reached, and the MOSES [46] algorithm by McDaid and Hurley using statistical model-based techniques. Concerning quality measures, extensions of the modularity metric for handling overlapping communities are described in [45, 50, 54].

2.4 Exceptional Model Mining

A general framework for multi-target quality functions in subgroup discovery is given by *exceptional model mining* [6, 41]: It tries to identify interesting patterns with respect to a local model derived from *a set* of attributes. The interestingness can be defined, e.g., by a significant deviation from a model that is derived from the total population or the respective complement set of instances within the population.

In general, a model consists of a specific *model class* and *model parameters* which depend on the values of the model attributes in the instances of the respective pattern cover. The quality measure q then determines the interestingness of a pattern according to its model parameters. Following [42], we outline some simple examples below, focusing on relations between pairs (correlation) and sets of variables (logistic regression):

- A relatively simple example for an exceptionality measure considers the task of identifying subgroups in which the correlation between two numeric attributes is especially strong, e.g., as measured by the Pearson correlation coefficient. This *correlation model class* has exactly one parameter, i.e., the correlation coefficient.

- Furthermore, using a *simple linear regression model*, we can compare the slopes of the regression lines of the subgroup to the general population or the subgroups' complement. This *simple linear regression model* shows the dependency between two numeric variables x and y: It is built by fitting a straight line in the two dimensional space by minimizing the squared residuals e_j of the model:

$$y_i = a + b \cdot x_i + e_j$$

The slope

$$b = \frac{\text{cov}(x, y)}{\text{var}(x)}$$

computed given the covariance $\text{cov}(x, y)$ of x and y, and the variance $\text{var}(x)$ of x can then be used for identifying interesting patterns [41].

- The *logistic regression model* is used for the classification of a binary target attribute $y \in T$ from a set of independent binary attributes $x_j \in T \setminus y, j = 1, \dots, |T| - 1$. The model is given by:

$$y = \frac{1}{1 + e^{-z}}, \ z = b_0 + \sum_j b_j x_j.$$

Interesting patterns are then those, for example, for which the model parameters b_j differ significantly from those derived from the total population.

Considering network structures, we can also adapt exceptional model mining to that setting. Essentially, it can be regarded as a description-oriented approach for assessing network structures, if the patterns are used to induce graphs or subgraphs. As we will discuss below, we can then also apply exceptional model mining for descriptive community detection, in essence combining subgroup discovery and community detection into a unified approach.

Below, we first outline a quality function for comparing graph structures that correspond to individual patterns (QAP). After that, we discuss quality functions used in community detection in order to assess subgraphs that are induced by some criterion, e.g., by a descriptive pattern.

For some notation, we follow the notions presented in [21]: As outlined above, the concept of a *community* intuitively describes a group C of individuals out of a population such that members of C are strongly "related" among each other but weakly "related" to individuals outside of C. By intuition, this relates, for example, to strongly connected groups of actors in social networks. This idea translates to communities as vertex sets $C \subseteq V$ of a graph $G = (V, E)$. To determine the amount of relatedness (or connectedness, and thus, the community quality of such a subset) several measures have been proposed.

For further concepts regarding our terminology and also the standard community quality functions outlined below, we follow the notation introduced in [21]: For a given undirected graph $G = (V, E)$ and a community $C \subseteq V$: $n := |V|$, let $m := |E|$, $n_C := |C|$, $m_C := |\{\{u, v\} \in E : u, v \in C\}|$—the number of *intra-edges* of C, and $\bar{m}_C := |\{\{u, v\} \in E : |\{u, v\} \cap C| = 1\}|$—the number of *inter-edges* of C. Here, it is also convenient to introduce an *inter-degree* for a node $u \in C$ (that depends on the choice of C) by $\bar{d}_C(u) := |\{\{u, v\} \in E : v \notin C\}|$, counting the number of edges between u and nodes outside of C, and $d(u) =:= |\{\{u, v\} \in E\}|$ is the degree of node u.

There is a wide range of different community evaluation functions $2^V \rightarrow \mathbb{R}$ for estimating the community quality. In the context of this paper, we focus on *maximizing* local quality functions for single communities (which are induced by specific patterns). Therefore, we consider the inverse of a quality measure in those cases, where the measure itself indicates higher quality by lower values.

- Concerning network structures, we can compare adjacency matrices induced by a specific pattern, see [7]. For the assessment we can apply, for example, the quadratic assignment procedure [39] (QAP): it is a standard approach for comparing network structures, e.g., using a graph correlation measure: For comparing two graphs G_1 and G_2, it estimates the correlation of the respective adjacency matrices M_1 and M_2 and tests that graph level statistic against a QAP null hypothesis [39].

  QAP compares the observed graph correlation of (G_1, G_2) to the distribution of the respective resulting correlation scores obtained on repeated random row and column permutations of the adjacency matrix of G_2. As a result, we obtain a correlation and a statistical significance level according to the randomized distribution scores.

  For deriving a quality measure based on QAP and graph correlation, we compare the reference matrix M_N and the matrix M_P for pattern P:

  $$q_Q(P) = \text{QAP}(M_N, M_P) = \frac{\text{cov}(M_N, M_P)}{\sqrt{\text{var}(M_N) \cdot \text{var}(M_P)}},$$

  where M_N is the transition matrix induced by some reference model (see [7, 24]), and M_P is the transition matrix induced by pattern P, *cov* indicates the covariance of the matrices, and $\text{var}(M) = \text{cov}(M, M)$ the variance.

  For an in-depth description of QAP, we refer to [39]. Furthermore, for the transition matrix, we refer to [23, 24] for more details on the matrix construction step.

- Regarding the quality of a subgraph induced by a pattern, we can adapt the well-known modularity measure to the idea of assessing the induced subgraph captured by a local pattern, i.e., a community pattern (with an associated subgroup description).

In general, the *modularity* MOD [51–53] of a graph clustering with k communities $C_1, \ldots, C_k \subseteq V$ focuses on the number of edges *within* a community and compares that with the *expected* such number given a null-model (i.e., a corresponding random graph where the node degrees of G are preserved). It is given by

$$\text{MOD} = \frac{1}{2m} \sum_{u,v \in V} \left(A_{u,v} - \frac{d(u)\, d(v)}{2m} \right) \delta(C(u), C(v)), \tag{3}$$

where $C(i)$ denotes for $i \in V$ the community to which node i belongs. $\delta(C(u), C(v))$ is the *Kronecker delta* symbol that equals 1 if $C(u) = C(v)$, and 0 otherwise. So, the *modularity* assesses the community quality of a graph partitioning, but can also be adapted to overlapping communities, e.g., [45, 50, 54] for considering the complete graph structure.

For exceptional model mining, however, we need to consider individual patterns. In order to focus on a subgraph induced by a pattern, the *modularity contribution* of a single community C in a *local context* (subgraph induced by the nodes contained in the community C) can then be computed [52, 54] as:

$$\text{MODL}(C) = \frac{1}{2m} \sum_{u,v \in C} \left(A_{u,v} - \frac{d(u)\, d(v)}{2m} \right),$$

yielding

$$\text{MODL}(C) = \frac{2m_C}{2m} - \sum_{u,v \in C} \frac{d(u)\, d(v)}{4m^2} = \frac{m_C}{m} - \sum_{u,v \in C} \frac{d(u)\, d(v)}{4m^2}.$$

- The *segregation index* SIDX [27] is another prominent measure from community detection. It focuses on the local contribution of the pattern, and compares the number of expected inter-edges to the number of observed inter-edges, normalized by the expectation:

$$\text{SIDX}(C) = \frac{E(\bar{m}_C) - \bar{m}_C}{E(\bar{m}_C)} = 1 - \frac{\bar{m}_C n(n-1)}{2mn_C(n - n_C)} \tag{4}$$

- Finally, the *Inverse Average-ODF (out-degree fraction)* IAODF [70] captures the basic intuition of a community regarding the contained vs. the outgoing edges discussed above. As another local measure, IAODF compares the number of *inter-edges* to the number of all edges of a community C, and averages this for the whole community by considering the fraction for each individual node:

$$\text{IAODF}(C) := 1 - \frac{1}{n_C} \sum_{u \in C} \frac{\bar{d}_C(u)}{d(u)} \tag{5}$$

3 Community Detection and Description

While the community detection methods described above only focus on the graph structure, richer graph representations, i.e., *attributed graphs*, enable approaches that specifically exploit the descriptive information of the labels assigned to nodes and/or edges of the graph. Nodes of a network representing users, for example, can be labeled with tags that the respective users utilized in social bookmarking systems, or nodes (denoting actors) can be labeled with properties of the latter. Then, *explicit descriptions* for the characterization of a community can be provided.

Concerning methods that focus on such descriptions in general, an approach for community detection using features identified by frequent pattern mining is presented in [1]; closed frequent patterns are derived and are then used for creating a social network model based on an entropy analysis. However, the network structure itself is not exploited. Similarly, [63] extracts subgraphs with common itemsets. Given a labeled graph, itemset-sharing subgraphs can then be enumerated. However, this approach also does not consider the density of graphs, nor any community measures.

Focusing on methods for generating *explicit descriptions connected with the graph structure*, we distinguish between two types of approaches: first, methods that mainly work on the graph structure but apply descriptive information for restricting the possible sets of communities; second, methods that mine descriptive patterns for obtaining community candidates evaluated using the graph structure. As a representative of the first type, the concepts of dense subgraphs and subspace clusters for mining cohesive patterns are combined in [49].

Starting with quasi-cliques, these are expanded until constraints regarding the description or the graph structure are violated. Similarly, [32] combines subspace clustering and dense subgraph mining, also interleaving quasi-clique and subspace construction. As an example for the second type outlined above, [28] proposes an approach for the problem of finding overlapping communities in graphs and social networks that aims at detecting the top-k communities such that the total edge density over all k communities is maximized. The three algorithmic variants proposed in [28] apply a greedy strategy for detecting dense subgroups, and restrict the result set of communities, such that each edge can belong to at most one community. This partitioning involves a global approach on the community quality. Furthermore, [64] study the correlation between attribute sets and the occurrence of dense subgraphs in large attributed graphs. The proposed method considers frequent attribute sets using an adapted frequent item mining technique, and identifies the top-k dense subgraphs induced by a particular attribute set, called structural correlation patterns. The DCM method presented in [57] includes a two-step process of community detection and community description. A heuristic approach is applied for discovering the top-k communities. Pool et al. utilize a special interestingness function which is based on counting outgoing edges of a community similar to the IAODF measure; for that, they also demonstrate the trend of a correlation with the modularity function.

Furthermore, the COMODO algorithm [21] that we summarize in the next section combines community detection and subgroup discovery resulting in a description-oriented approach. By specifying a standard quality function the quality of the communities to discover can be estimated. Then, this quality function can be specifically selected according to the analysis task.

4 Community Detection Using Exceptional Model Mining

For providing both structurally valid and interpretable communities we utilize the graph structure as well as additional descriptive features of the nodes. Hence, we identify communities as sets of nodes together with a *description* composed of the nodes' features. Such a *community pattern* then provides an intuitive description of the community, e.g., by an easily interpretable conjunction of attribute-value pairs. Basically, we aim at identifying communities according to standard community quality measures. Below, we first provide an algorithmic overview on the approach and summarize exemplary evaluation results. After that, we sketch the application of the algorithm for community detection on dynamic networks, i.e., for identifying exceptional sequential patterns.

4.1 COMODO: *Description-Oriented Community Detection*

Below, we summarize the COMODO algorithm presented in [21]: It focuses on *description-oriented community detection* using subgroup discovery, and aims at discovering the top-k communities (described by community patterns). The method is based on an adapted subgroup discovery approach [12, 42], and also tackles typical problems that are not addressed by standard approaches for community detection such as pathological cases like small community sizes. COMODO utilizes optimistic estimates [31, 67], which are efficient to compute, in order to prune the search space significantly. For that, a number of standard community evaluation functions have been applied using optimistic estimates for an efficient approach.

4.1.1 Algorithmic Overview

COMODO utilizes both the graph structure and descriptive information of the attributed graph. This information is contained in two data structures: The graph structure is encoded in graph G while the attribute information is contained in database D describing the respective attribute values of each node. In a preprocessing step, we merge these data sources. Since the communities considered in our approach do not contain isolated nodes, we can describe them as sets of edges. We transform the data (of the given graph G and the database D containing the nodes' descriptive information) into a new dataset focusing on the edges of the graph G:

Each data record in the new dataset represents an edge between two nodes. The attribute values of each such data record are the common attributes of the edge's two nodes. For a more detailed description, we refer to [21].

COMODO utilizes an extended FP-tree (frequent pattern tree) structure inspired by the FP-growth algorithm, which compiles the data in a convenient prefix pattern tree structure for mining frequent itemsets, see [2] for a detailed description. Our adapted tree structure is called the *community pattern tree* (CP-tree) that allows to efficiently traverse the solution space. The tree is built in two scans of the graph dataset and is then mined in a recursive divide-and-conquer manner, see [9, 42] for more details. In the main algorithmic procedure of COMODO, patterns containing only one basic pattern are mined first. Then, patterns conditioned on the occurrence of a (prefixed) complex pattern (as a set of basic patterns, chosen in the previous recursion step) are considered recursively. For more algorithmic details, we refer to [21]. As described there, we can apply standard quality functions efficiently using optimistic estimates, e.g., for the *modularity* or the *segregation index*, see [21] for more details.

4.1.2   Illustrative Evaluation Results

Below, we present illustrative evaluation results [21] considering the efficiency of the applied optimistic estimates, and the validity of the obtained patterns. For that, we compared the total number of search steps, that is community allocations that are considered by the COMODO algorithm, with no optimistic estimate pruning to optimistic estimate pruning using different community quality measures. Additionally, we measured the impact of using different minimal community size thresholds. Some results are shown in Fig. 1 for the BibSonomy click graph for

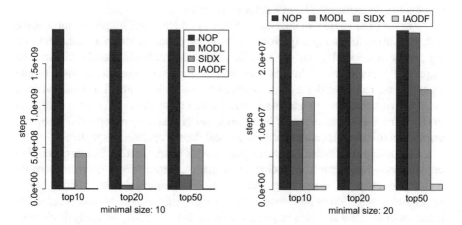

Fig. 1 Runtime performance of COMODO on the BibSonomy click graph, see [21] for more details: search steps with no optimistic estimate pruning (*NOP*) vs. community quality functions with optimistic estimate pruning: MODL (Local modularity), SIDX (Segregation Index), and IAODF (Inverse Average-ODF), for minimal size thresholds $\tau_n = 10, 20$

$k = 10, 20, 50$ and minimal size thresholds $\tau_n = 10, 20$. We consider a number of standard community quality functions, that is, the *segregation index*, the *Inverse Average-ODF*, and the *modularity*.

The large, exponential search space can be exemplified, e.g., for the click graph with a total of about $2 \cdot 10^{10}$ search steps for a minimal community size threshold $\tau_n = 10$. The results demonstrate the effectiveness of the proposed descriptive mining approach applying the presented optimistic estimates. The implemented pruning scheme makes the approach scalable for larger datasets, especially when the local modularity quality function is chosen to assess the communities' quality. Concerning the validity of the patterns, we focused on structural properties of the patterns and the subgraphs induced by the respective community patterns. We applied the significance test described in [38] for testing the statistical significance of the density of a discovered subgraph. Furthermore, we compared COMODO to three baseline community detection algorithms [30, 46, 57], where COMODO consistently shows a significantly better performance concerning validity and description length; for more details, we refer to [21].

4.2 Sequential Pattern Analysis: Detecting Exceptional Link Trails

In addition to static community detection, we can also consider temporal aspects, i.e., focusing on sequences of states or events which can be applied for a variety of analysis ranging from the analysis of human behavior [23] to industrial applications [24]. In an extended modeling approach, we can map transitions between states to a weighted network, according to a first-order Markov chain model. Below, we outline an approach for detecting exceptional sequential link trails captured by community patterns, see [7] for a detailed description.

As before, our subject of analysis is given by an attributed graph that models the link trails in the following way: Nodes of the graph denote actors of a social network, e.g., users of a social system or locations in a location-based social network. The edges of the graph model the links between the nodes (as transitions). As a simple example, we can consider a set of users and a set of locations. Each user visits a sequence of locations—in a location-based social network. Then, we are interested in modeling these sequences (of locations), and in detecting exceptional groups of transitions (between locations) w.r.t. users and their properties, respectively.

At a music event festival, for example, possible characterizing factors describing certain users groups could be specific music genres. Here, exceptional patterns could include, for example, users being interested in *rock music* and *dance* visiting only a very specific selection of performances in characteristic sequences, compared to the behavior of all users and their sequential link trails. Essentially, we apply descriptive community detection (e.g., using COMODO) on the attributed graph, where the edges indicate transitions between states according to a first-order Markov chain modeling approach [44, 65].

4.2.1 Modeling

For our attributed graph model, we label the links according to the descriptive information of the sequential trail. Then, we identify exceptional community patterns based on the labels and structure of the contained links using exceptional model mining. In particular, we assess a pattern capturing a set of nodes that model the state space of the respective transitions.

For constructing a reference model, we construct transition matrices corresponding to the *observed data*. For those observed sequences we can simply construct transition matrices counting the transitions between the individual states. We construct an according matrix M^N with $m_{ij}^N = |\text{suc}(i,j)|$, where $\text{suc}(i,j)$ denotes the successive sequences from state i to state j contained in the sequence.

A community pattern P induces a subgraph (community) C_P given a set of labels P, selecting all links that are covered, i.e., that share a label contained in P. Then, all transitions in the matrix M^N are selected (corresponding to a set of links of the network) that are covered by the pattern P. Using that, we construct an according transition pattern matrix M^P based on the respective counts of the covered transitions. Intuitively, the matrix M^P can then be regarded as some kind of "projection" of matrix M^N given the pattern P using our modeling approach. In the simplest case, we can just transfer the weighted links of the subgraph C_P. For identifying exceptional models (M_P induced by P) we can then apply, e.g., the QAP quality function $q_Q(P) = \text{QAP}(M_N, M_P)$ introduced above.

4.2.2 Results

For some illustrative results (see [7] for more details), we utilized data from the EveryAware[1] project, e.g., [19]. Specifically, we focused on collectively organized noise measurements collected using the *WideNoise Plus* application between December 14, 2011 and June 6, 2014, see [20] for more details. *WideNoise Plus* allows the collection of noise measurements using smartphones. It includes sensor data from the microphone given as noise level in dB(A), the location from the GPS-, GSM-, and WLAN-sensor represented as latitude and longitude coordinate, as well as a timestamp. In addition, tags can be assigned to the recording. We collected data from all around the world using iOS and Android devices.

In total, the applied dataset contains 6069 data records, i.e., noise measurements of 635 users (i.e., 635 trails, with an average trail length of about 10) and 2009 distinct tags. Table 1 shows exemplary exceptional conforming and deviating patterns using q_Q as quality measure. In addition, it shows the sizes of the covered subsets. From a qualitative point of view, the patterns shown in the table are intuitive

---

[1]http://www.everyaware.eu.

Table 1 Illustrative
exceptional
conforming/deviating
community patterns for
WideNoise Plus

#	q_Q	Size	Description
1	0.94	5078	Traffic
2	0.89	3990	Car
3	0.76	3326	Noise
4	0.43	707	Bird ∧ courtyard
5	0.24	600	Background ∧ quiet

Patterns #1–#3 tend rather to conform to
the reference model (especially #1 and
#2), while patterns #4–#5 (increasingly)
show a deviating behavior

to interpret and also tend to conform to our expectations concerning the reference
behavior of the dataset, where we can clearly identify deviations concerning noisy
and relatively quiet environments.

5 Conclusions

In this paper, we have presented an organized view on descriptive community
detection. Specifically, we described subgroup discovery for compositional network
analysis concerning properties of the actors, with extensions to the analysis of
complex target concepts like correlations between a set of variables, or dense
subgraphs—captured by exceptional model mining approaches. Then, this directly
extends to community detection on attributed graphs. In particular, we summarized
the COMODO algorithm that combines community detection and exceptional
model mining, resulting in a description-oriented approach for community analytics.
We furthermore sketched an extension to dynamic data, considering sequential patterns capturing exceptional sequential link trails. This adds one further dimension to
the descriptive approaches, by considering by static as well as dynamic phenomena,
and enables the modeling and investigation of complex analysis tasks.

For future work, we aim to extend the analysis towards further time-oriented
representations, e.g., considering sequences of graphs, and the evolution of communities, e.g., [33, 34]. Also, we aim to integrate and exploit methods for generating
descriptions and the respective relations in link analytics, e.g., in link predic-
tion [60–62] on multiplex networks. Then, besides the detection of communities,
also their analysis and assessment in the form of descriptive patterns is highly
relevant, e.g., [11, 15, 17, 18] also concerning their semantic grounding [47, 48], and
integration into explanation-aware approaches [16, 25, 59]. Furthermore, developing
scalable methods for enabling such approaches for large and complex datasets,
e.g., [22, 42] is another interesting direction for future work.

References

1. Adnan, M., Alhajj, R., Rokne, J.: Identifying social communities by frequent pattern mining. In: Proc. 13th Intl. Conf. Information Visualisation, pp. 413–418. IEEE Computer Society, Washington, DC (2009)
2. Agrawal, R., Srikant, R.: Fast algorithms for mining association rules. In: Bocca, J.B., Jarke, M., Zaniolo, C. (eds.) Proc. 20th Int. Conf. Very Large Data Bases (VLDB), pp. 487–499. Morgan Kaufmann, San Francisco (1994)
3. Agresti, A.: An Introduction to Categorical Data Analysis. Wiley, Hoboken (2007)
4. Atzmueller, M.: Data mining on social interaction networks. J. Data Min. Digit. Humanit. **1**, pp. 1–34 (2014)
5. Atzmueller, M.: Subgroup and community analytics on attributed graphs. In: Kuznetsov, S.O., Missaoui, R., Obiedkov, S. (eds.) Proceedings of the International Workshop on Social Network Analysis Using Formal Concept Analysis (SNAFCA-2015), CEUR-WS, vol. 1534 (2015)
6. Atzmueller, M.: Subgroup discovery – advanced review. WIREs Data Min. Knowl. Discov. **5**(1), 35–49 (2015)
7. Atzmueller, M.: Detecting community patterns capturing exceptional link trails. In: Proceedings of the IEEE/ACM ASONAM. IEEE Press, Boston, MA (2016)
8. Atzmueller, M.: Local exceptionality detection on social interaction networks. In: Proceedings of the ECML-PKDD 2016: European Conference on Machine Learning and Principles and Practice of Knowledge Discovery in Databases. Springer, Berlin (2016)
9. Atzmueller, M., Lemmerich, F.: Fast subgroup discovery for continuous target concepts. In: Proceedings of the International Symposium on Methodologies for Intelligent Systems. LNCS, vol. 5722, pp. 1–15. Springer, Heidelberg (2009)
10. Atzmueller, M., Lemmerich, F.: VIKAMINE - open-source subgroup discovery, pattern mining, and analytics. In: Proceedings of the European Conference on Machine Learning and Principles and Practice of Knowledge Discovery in Databases. Springer, Heidelberg (2012)
11. Atzmueller, M., Lemmerich, F.: Exploratory pattern mining on social media using geo-references and social tagging information. Int. J. Web Sci. **2**(1/2), 80–112 (2013)
12. Atzmueller, M., Mitzlaff, F.: Efficient descriptive community mining. In: Proceedings of the 24th International FLAIRS Conference, pp. 459–464. AAAI Press, Palo Alto, CA (2011)
13. Atzmueller, M., Puppe, F.: Semi-automatic visual subgroup mining using VIKAMINE. J. Univers. Comput. Sci. **11**(11), 1752–1765 (2005)
14. Atzmueller, M., Puppe, F.: SD-Map - a fast algorithm for exhaustive subgroup discovery. In: Proceedings of the European Conference on Principles and Practice of Knowledge Discovery in Databases (PKDD), pp. 6–17. Springer, Heidelberg (2006)
15. Atzmueller, M., Puppe, F.: A case-based approach for characterization and analysis of subgroup patterns. J. Appl. Intell. **28**(3), 210–221 (2008)
16. Atzmueller, M., Roth-Berghofer, T.: The mining and analysis continuum of explaining uncovered. In: Proceedings of the 30th SGAI International Conference on Artificial Intelligence (AI-2010) (2010)
17. Atzmueller, M., Baumeister, J., Hemsing, A., Richter, E.J., Puppe, F.: Subgroup mining for interactive knowledge refinement. In: Proceedings of the 10th Conference on Artificial Intelligence in Medicine (AIME 05). LNAI, vol. 3581, pp. 453–462. Springer, Heidelberg (2005)
18. Atzmueller, M., Baumeister, J., Puppe, F.: Introspective subgroup analysis for interactive knowledge refinement. In: Proceedings of the 19th International Florida Artificial Intelligence Research Society Conference 2006 (FLAIRS-2006), pp. 402–407. AAAI Press, Palo Alto, CA (2006)

19. Atzmueller, M., Becker, M., Kibanov, M., Scholz, C., Doerfel, S., Hotho, A., Macek, B.E., Mitzlaff, F., Mueller, J., Stumme, G.: Ubicon and its applications for ubiquitous social computing. New Rev. Hypermedia Multimed. **20**(1), 53–77 (2014)
20. Atzmueller, M., Mueller, J., Becker, M.: Exploratory subgroup analytics on ubiquitous data. In: Mining, Modeling and Recommending 'Things' in Social Media. LNAI, vol. 8940. Springer, Heidelberg (2015)
21. Atzmueller, M., Doerfel, S., Mitzlaff, F.: Description-oriented community detection using exhaustive subgroup discovery. Inf. Sci. **329**, 965–984 (2016)
22. Atzmueller, M., Mollenhauer, D., Schmidt, A.: Big data analytics using local exceptionality detection. In: Enterprise Big Data Engineering, Analytics, and Management. IGI Global, Hershey, PA (2016)
23. Atzmueller, M., Schmidt, A., Kibanov, M.: DASHTrails: an approach for modeling and analysis of distribution-adapted sequential hypotheses and trails. In: Proceedings of the WWW 2016 (Companion), IW3C2/ACM (2016)
24. Atzmueller, M., Schmidt, A., Kloepper, B., Arnu, D.: HypGraphs: an approach for modeling and comparing graph-based and sequential hypotheses. In: Proceedings of the ECML-PKDD Workshop on New Frontiers in Mining Complex Patterns (NFMCP), Riva del Garda (2016)
25. Clancey, W.J.: The epistemology of a rule-based expert system: a framework for explanation. Artif. Intell. **20**, 215–251 (1983)
26. Fortunato, S.: Community detection in graphs. Phys. Rep. **486**(3–5), 75–174 (2010)
27. Freeman, L.: Segregation in social networks. Sociol. Methods Res. **6**(4), 411 (1978)
28. Galbrun, E., Gionis, A., Tatti, N.: Overlapping community detection in labeled graphs. Data Min. Knowl. Discov. **28**(5–6), 1586–1610 (2014)
29. Geng, L., Hamilton, H.J.: Interestingness measures for data mining: a survey. ACM Comput. Surv. **38**(3), Article No. 9 (2006)
30. Gregory, S.: Finding overlapping communities in networks by label propagation. New J. Phys. **12**, 103018 (2010)
31. Grosskreutz, H., Rüping, S., Wrobel, S.: Tight optimistic estimates for fast subgroup discovery. In: Proceedings of the European Conference on Machine Learning and Principles and Practice of Knowledge Discovery in Databases. LNCS, vol. 5211, pp. 440–456. Springer, Heidelberg (2008)
32. Günnemann, S., Färber, I., Boden, B., Seidl, T.: GAMer: a synthesis of subspace clustering and dense subgraph mining. In: Knowledge and Information Systems. Springer, London (2013)
33. Kibanov, M., Atzmueller, M., Scholz, C., Stumme, G.: Temporal evolution of contacts and communities in networks of face-to-face human interactions. Sci. China **57**, 1–17 (2014)
34. Kibanov, M., Atzmueller, M., Illig, J., Scholz, C., Barrat, A., Cattuto, C., Stumme, G.: Is web content a good proxy for real-life interaction? A case study considering online and offline interactions of computer scientists. In: Proceedings of the IEEE/ACM International Conference on Advances in Social Networks Analysis and Mining (ASONAM). IEEE Press, Boston, MA (2015)
35. Klösgen, W.: Explora: a multipattern and multistrategy discovery assistant. In: Fayyad, U.M., Piatetsky-Shapiro, G., Smyth, P., Uthurusamy, R. (eds.) Advances in Knowledge Discovery and Data Mining, pp. 249–271. AAAI Press, Menlo Park (1996)
36. Klösgen, W.: 16.3: subgroup discovery. In: Handbook of Data Mining and Knowledge Discovery. Oxford University Press, New York (2002)
37. Klösgen, W.: 5.2: subgroup patterns. In: Handbook of Data Mining and Knowledge Discovery. Oxford University Press, New York (2002)
38. Koyuturk, M., Szpankowski, W., Grama, A.: Assessing significance of connectivity and conservation in protein interaction networks. J. Comput. Biol. **14**(6), 747–764 (2007)
39. Krackhardt, D.: QAP partialling as a test of spuriousness. Soc. Netw. **9**, 171–186 (1987)
40. Lancichinetti, A., Fortunato, S., Kertsz, J.: Detecting the overlapping and hierarchical community structure in complex networks. New J. Phys. **11**(3), 033015 (2009)

41. Leman, D., Feelders, A., Knobbe, A.: Exceptional model mining. In: Proceedings of the European Conference on Machine Learning and Principles and Practice of Knowledge Discovery in Databases. Lecture Notes in Computer Science, vol. 5212, pp. 1–16. Springer, Heidelberg (2008)
42. Lemmerich, F., Becker, M., Atzmueller, M.: Generic pattern trees for exhaustive exceptional model mining. In: Proceedings of the European Conference on Machine Learning and Principles and Practice of Knowledge Discovery in Databases. Springer, Heidelberg (2012)
43. Lemmerich, F., Atzmueller, M., Puppe, F.: Fast exhaustive subgroup discovery with numerical target concepts. Data Min. Knowl. Discov. **30**, 711–762 (2016)
44. Lempel, R., Moran, S.: The stochastic approach for link-structure analysis (SALSA) and the TKC effect. Comput. Netw. **33**(1), 387–401 (2000)
45. Lin, Y.R., Chi, Y., Zhu, S., Sundaram, H., Tseng, B.L.: Analyzing communities and their evolutions in dynamic social networks. ACM Trans. Knowl. Discov. Data **3**, 8:1–8:31 (2009)
46. McDaid, A., Hurley, N.: Detecting highly overlapping communities with model-based overlapping seed expansion. In: Proceedings of the International Conference on Advances in Social Networks Analysis and Mining, ASONAM, pp. 112–119. IEEE Computer Society, Washington, DC (2010)
47. Mitzlaff, F., Atzmueller, M., Stumme, G., Hotho, A.: Semantics of user interaction in social media. In: Ghoshal, G., Poncela-Casasnovas, J., Tolksdorf, R. (eds.) Complex Networks IV. Studies in Computational Intelligence, vol. 476. Springer, Heidelberg (2013)
48. Mitzlaff, F., Atzmueller, M., Hotho, A., Stumme, G.: The social distributional hypothesis. J. Soc. Netw. Anal. Min. **4**, 216 (2014)
49. Moser, F., Colak, R., Rafiey, A., Ester, M.: Mining cohesive patterns from graphs with feature vectors. In: SDM, SIAM, vol. 9, pp. 593–604 (2009)
50. Muff, S., Rao, F., Caflisch, A.: Local modularity measure for network clusterizations. Phys. Rev. E Stat. Nonlinear Matter Phys. **72**(5), 056107 (2005)
51. Newman, M.E.J.: Detecting community structure in networks. Eur. Phys. J. **38**, 321–330 (2004)
52. Newman, M.E.J.: Modularity and community structure in networks. Proc. Natl. Acad. Sci. **103**(23), 8577–8582 (2006)
53. Newman, M.E.J., Girvan, M.: Finding and evaluating community structure in networks. Phys. Rev. E Stat. Nonlin Soft Matter Phys. **69**(2), 1–15 (2004)
54. Nicosia, V., Mangioni, G., Carchiolo, V., Malgeri, M.: Extending the definition of modularity to directed graphs with overlapping communities. J. Stat. Mech. **2009**, 03024 (2009)
55. Palla, G., Dernyi, I., Farkas, I., Vicsek, T.: Uncovering the overlapping community structure of complex networks in nature and society. Nature **435**(7043), 814–818 (2005)
56. Palla, G., Farkas, I.J., Pollner, P., Derenyi, I., Vicsek, T.: Directed network modules. New J. Phys. **9**(6), 186 (2007)
57. Pool, S., Bonchi, F., van Leeuwen, M.: Description-driven community detection. Trans. Intell. Syst. Technol. **5**(2), 1–21 (2014)
58. Raghavan, U., Albert, R., Kumara, S.: Near linear time algorithm to detect community structures in large-scale networks. Phys. Rev. E **76**, 036106 (2007)
59. Roth-Berghofer, T.R., Cassens, J.: Mapping goals and kinds of explanations to the knowledge containers of case-based reasoning systems. In: Muñoz-Avila, H., Ricci, F. (eds.) Case-Based Reasoning Research and Development, 6th International Conference on Case-Based Reasoning, ICCBR 2005, Chicago, IL, USA, August 2005, Proceedings. Lecture Notes in Artificial Intelligence, vol. 3620, pp. 451–464. Springer, Heidelberg (2005)
60. Scholz, C., Atzmueller, M., Stumme, G.: On the predictability of human contacts: influence factors and the strength of stronger ties. In: Proceedings of the 4th ASE/IEEE International Conference on Social Computing (SocialCom). IEEE Computer Society, Boston, MA (2012)
61. Scholz, C., Atzmueller, M., Barrat, A., Cattuto, C., Stumme, G.: New Insights and Methods For Predicting Face-To-Face Contacts. In: Kiciman E, Ellison NB, Hogan B, Resnick P, Soboroff I (eds) Proc. International AAAI Conference on Weblogs and Social Media. AAAI Press, Palo Alto, CA (2013)

62. Scholz, C., Atzmueller, M., Kibanov, M., Stumme, G.: How do people link? Analysis of contact structures in human face-to-face proximity networks. In: Proc. ASONAM 2013. ACM Press, New York, NY (2013)
63. Sese, J., Seki, M., Fukuzaki, M.: Mining networks with shared items. In: Proceedings of the 19th ACM International Conference on Information and Knowledge Management, pp. 1681–1684. ACM, New York, NY (2010)
64. Silva, A., Meira Jr., W., Zaki, M.J.: Mining attribute-structure correlated patterns in large attributed graphs. Proc VLDB Endowment 5(5), 466–477 (2012)
65. Singer, P., Helic, D., Taraghi, B., Strohmaier, M.: Detecting memory and structure in human navigation patterns using Markov chain models of varying order. PLoS One 9(7), e102070 (2014)
66. Wasserman, S., Faust, K.: Social Network Analysis: Methods and Applications. Structural Analysis in the Social Sciences, vol. 8, 1st edn. Cambridge University Press, Cambridge (1994)
67. Wrobel, S.: An algorithm for multi-relational discovery of subgroups. In: Proceedings of the 1st European Symposium on Principles of Data Mining and Knowledge Discovery, pp. 78–87. Springer, Heidelberg (1997)
68. Xie, J., Szymanski, B.K.: LabelRank: a stabilized label propagation algorithm for community detection in networks. In: Proceedings of the IEEE Network Science Workshop, West Point, NY (2013)
69. Xie, J., Kelley, S., Szymanski, B.K.: Overlapping community detection in networks: the state-of-the-art and comparative study. ACM Comput. Surv. 45(4), 43:1–43:35 (2013)
70. Yang, J., Leskovec, J.: Defining and evaluating network communities based on ground-truth. In: Proceedings of the ACM SIGKDD Workshop on Mining Data Semantics, MDS '12, pp. 3:1–3:8. ACM, New York, NY (2012)

Multimodal Clustering for Community Detection

Dmitry I. Ignatov, Alexander Semenov, Daria Komissarova, and Dmitry V. Gnatyshak

1 Introduction

Online social networking services generate massive amounts of data, which can become a valuable source for guiding Internet advertisement efforts or provide sociological insights. Each registered user has a network of friends as well as specific profile features. These profile features describe the user's tastes, preferences, the groups he or she belongs to, etc. Social Network Analysis (SNA) is a popular research field in which methods are developed for analysing one-mode networks, like friend-to-friend,[1] two-mode or affiliation networks [57, 60, 69], three-mode [10, 20, 38, 46, 66], and even multi-mode dynamic networks [75, 76, 81, 89]. By multi-mode networks we mean namely such networks where actors can be related with other types of entities by edges like those between users and their interests in two-mode case or by hyperedges like those relating users, tags, and resources in three-mode case; sometimes such networks are called heterogeneous since different types of nodes are involved [48]. We focus on the subfield of bicommunity identification and its higher order extensions. Thus, in particular, we present tri- and tetracommunities examples extracted from real data. For one-mode case a reader may refer to an extensive survey on community detection [21].

---

[1] www.https://en.wikipedia.org/wiki/Friend-to-friend.

D.I. Ignatov (✉) • D. Komissarova • D.V. Gnatyshak
National Research University Higher School of Economics, Moscow, Russia
e-mail: dignatov@hse.ru; komissarovadaria93@gmail.com; dgnatyshak@hse.ru

A. Semenov
National Research University Higher School of Economics, Moscow, Russia

Mobile TeleSystems PJSC, Moscow, Russia
e-mail: SemenoffAlex@gmail.com

© Springer International Publishing AG 2017
R. Missaoui et al. (eds.), *Formal Concept Analysis of Social Networks*,
Lecture Notes in Social Networks, DOI 10.1007/978-3-319-64167-6_4

The notion of community in SNA and Complex Networks is closely related to the notion of cluster in Data Analysis [3, 21]. There is the main issue in both disciplines: what is a common definition of community and what is a common definition of cluster? On the one hand, it is clear that actors from the same community should be similar as well as objects in one cluster; on the other hand, these actors (or objects) should be less similar to actors (or objects) from another community (or cluster). This general idea allows a variety of definitions suitable for concrete purposes in both domains [3, 21, 63].

There is a large amount of network data that can be represented as bipartite or tripartite graphs. Standard techniques for community detection in two-mode networks like "maximal bicliques search" return a huge number of patterns (in the worst case exponential w.r.t. the input size) [56, 77]. Moreover, not all members of such bicommunities should be related to the same items, for example, exactly the same vocabulary used by each member in case of epistemic communities. Therefore we need some relaxation of the biclique notion as well as appropriate interestingness measures and constraints for mining and filtering such "relaxed" biclique communities.

Applied lattice theory provides us with the definition of formal concept [27], which is closely related to maximal biclique in a bipartite graph; formal concepts and concept lattices (or Galois lattices) are widely known in the social network analysis community (see, e.g., [19, 23, 24, 65, 77, 86]). However, these methods are overly rigid for analysing large amounts of data resulting in a huge number of concepts even if their computation is feasible.

A concept-based bicluster (or object-attribute bicluster) [37] is a scalable approximation of a formal concept (maximal biclique in a bipartite graph). The advantages of concept-based biclustering are

1. Less number of patterns to analyse (no more than the number of edges in the original network);
2. Less computational time (polynomial vs exponential);
3. Tolerance to missing (object, attribute) pairs;
4. Filtering of biclusters (communities) by density threshold.

In general, the method of biclustering dates back to the seminal work of Hartigan on the so-called direct clustering [31], where clusters of objects may appear sharing only a subset of attributes. The term biclustering was introduced later in the book of Mirkin [63]:

> The term biclustering refers to simultaneous clustering of both row and column sets in a data matrix. Biclustering addresses the problems of aggregate representation of the basic features of interrelation between rows and columns as expressed in the data.

Following this terminology, formal concepts can be considered as maximal inclusion biclusters of constant values in binary data [49], whereas their relaxations tolerant to missing object-attribute pairs can be called object-attribute biclusters [37, 39].

There are several successful attempts to mine two-mode [51, 78], three-mode [46], and even four-mode communities [47] by means of Formal Concept Analysis.

For analysing three-mode network data like folksonomies [83] we have also proposed a scalable triclustering technique [40, 45].

These studies for higher-mode cases were enabled by the previous introduction of the so-called triconcepts by Lehman and Wille [58, 87]; a formal triconcept consists of three components: extent (objects), intent (attributes), and modus (conditions under which an object has an attribute). It is a matter of curiosity, but such triconcepts had been used for analysing triadic data in social cognition studies [52] before their formal introduction. Later, a polyadic (or multimodal) extension of FCA was introduced in [85].

Previously, we have introduced a pseudo-triclustering technique for tagging groups of users by their common interests [28]. This approach differs from traditional triclustering methods because it relies on the extraction of biclusters from two separate object-attribute tables and rather belongs to methods for analysing multi-relational networks. Here we investigate applicability of biclustering and triclustering (as well as n-clustering, its higher-mode extension) to community detection in two-, three-, and higher-mode networks directly.

The remainder of the paper is organised as follows. In Sect. 2, we introduce basic notions of Formal Concept Analysis. Section 3 describes object-attribute biclustering and its direct generalisations to higher dimensions. Section 4 briefly discusses a variety of quality measures used in clustering, FCA, and SNA domains and their interrelation with multimodal clustering. In Sect. 5, we describe datasets which we have chosen to illustrate the performance of the approach. We present the results obtained during experiments on these datasets in Sect. 6. Related work is discussed in Sect. 7, while Sect. 8 concludes our paper and describes some interesting directions for future research.

2   Basic Definitions

2.1   Formal Concept Analysis

A *formal context* in FCA [27] is a triple $\mathbb{K} = (G, M, I)$, where G is a *set of objects*, M is a *set of attributes*, and the relation $I \subseteq G \times M$ shows which object possesses which attribute. For any $A \subseteq G$ and $B \subseteq M$ one can define *Galois operators*:

$$A' = \{m \in M \mid gIm \text{ for all } g \in A\},$$
$$B' = \{g \in G \mid gIm \text{ for all } m \in B\}. \qquad (1)$$

The operator $''$ (applying the operator $'$ twice) is a *closure operator*: it is idempotent ($A'''' = A''$), monotone ($A \subseteq B$ implies $A'' \subseteq B''$), and extensive ($A \subseteq A''$). The set of objects $A \subseteq G$ such that $A'' = A$ is called closed. Similar properties are valid for closed attribute sets, subsets of a set M. A pair (A, B) such that $A \subseteq G$, $B \subseteq M$, $A' = B$, and $B' = A$, is called a *formal concept* of a context $\mathbb{K}$. The sets A and B are closed and called *extent* and *intent* of a formal concept

(A, B), respectively. For the set of objects A the set of their common attributes A' describes the similarity of objects of the set A, and the closed set A'' is a cluster of similar objects (with the set of common attributes A'). The relation "to be a more general concept" is defined as follows: $(A, B) \geq (C, D)$ iff $A \supseteq C$. The concepts of a formal context $\mathbb{K} = (G, M, I)$ ordered by extensions inclusion form a lattice, which is called *concept lattice*. For its visualisation *line diagrams* (Hasse diagrams) can be used, i.e. the cover graph of the relation "to be a more general concept". In the worst case (Boolean lattice) the number of concepts is equal to $2^{\{\min |G|, |M|\}}$, thus, for large contexts, to make application of FCA machinery tractable the data should be sparse. Moreover, one can use different ways of filtering formal concepts (for example, choosing concepts by their stability index or extent size).

Let us consider a formal context $\mathbb{K}$ that consists of four objects, persons (Alex, Mike, Kate, David), four attributes, books (Romeo and Juliet by William Shakespeare, The Puppet Masters by Robert A. Heinlein, Ubik by Philip K. Dick, and Ivanhoe by Walter Scott), and incidence relation showing which person which book read or liked.

$\mathbb{K}$	Romeo and Juliet	The Puppets Masters	Ubik	Ivanhoe
Kate	×			×
Mike	×		×	
Alex		×	×	
David		×	×	×

There are nine concepts there. For example,
$C_1 = (\{Kate, Mike\}, \{Romeo\ and\ Juliet\})$
$C_2 = (\{Alex, David\}, \{The\ Puppet\ Masters, Ubik\})$
$C_3 = (\{Kate, David\}, \{Ivanhoe\})$.
Note that the pair of sets $(A, B) = (\{Alex, David\}, \{Ubik\})$ does not form a formal concept since we can enlarge its extent by one more object Mike to fulfill $(A \cup \{Mike\})' = B$ and $B' = A \cup \{Mike\}$. So, $C_4 = (\{MIke, Alex, David\}, \{Ubik\})$ is a formal concept. The corresponding bipartite graph is shown in Fig. 1 along with the biclique formed by elements of concept C_2.

Fig. 1 Two-mode network of readers and its community of Sci-Fi readers (*shaded*)

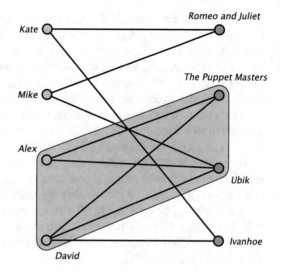

From SNA viewpoint, if we assume that an OA-bicluster (*event′, actor′*) is a found community, we are looking for a pair (*actor, event*) in an input network, where this actor participated in all of the events typical for the community, while the chosen event is typical for all the members of that community.

3   Higher-Order Extensions of FCA and Multimodal Clustering

3.1   Triadic and Polyadic FCA

For convenience, a *triadic context* is denoted by (X_1, X_2, X_3, Y). A triadic context $\mathbb{K} = (X_1, X_2, X_3, Y)$ gives rise to the following dyadic contexts:

$$\mathbb{K}^{(1)} = (X_1, X_2 \times X_3, Y^{(1)}), \quad \mathbb{K}^{(2)} = (X_2, X_1 \times X_3, Y^{(2)}), \quad \mathbb{K}^{(3)} = (X_3, X_1 \times X_2, Y^{(3)}),$$

where $gY^{(1)}(m, b) :\Leftrightarrow mY^{(2)}(g, b) :\Leftrightarrow bY^{(3)}(g, m) :\Leftrightarrow (g, m, b) \in Y$. The *derivation operators* (primes or concept-forming operators) induced by $\mathbb{K}^{(i)}$ are denoted by $(.)^{(i)}$. For each induced dyadic context we have two kinds of such derivation operators. That is, for $\{i, j, k\} = \{1, 2, 3\}$ with $j < k$ and for $Z \subseteq X_i$ and $W \subseteq X_j \times X_k$, the (i)-derivation operators are defined by:

$$Z \mapsto Z^{(i)} = \{(x_j, x_k) \in X_j \times X_k | x_i, x_j, x_k \text{ are related by Y for all } x_i \in Z\},$$

$$W \mapsto W^{(i)} = \{x_i \in X_i | x_i, x_j, x_k \text{ are related by Y for all } (x_j, x_k) \in W\}.$$

Formally, a *triadic concept* of a triadic context $\mathbb{K} = (X_1, X_2, X_3, Y)$ is a triple (A_1, A_2, A_3) of $A_1 \subseteq X_1, A_2 \subseteq X_2, A_3 \subseteq X_3$, such that for every $\{i, j, k\} = \{1, 2, 3\}$ with $j < k$ we have $(A_j \times A_k)^{(i)} = A_i$. For a certain triadic concept (A_1, A_2, A_3), the components A_1, A_2, and A_3 are called the *extent*, the *intent*, and the *modus* of (A_1, A_2, A_3). Since a tricontext $\mathbb{K} = (X_1, X_2, X_3, Y)$ can be interpreted as a three-dimensional cross table, according to our definition, under suitable permutations of rows, columns, and layers of this cross table, the triadic concept (A_1, A_2, A_3) is interpreted as a maximal cuboid full of crosses. The set of all triadic concepts of $\mathbb{K} = (X_1, X_2, X_3, Y)$ is denoted by $\mathfrak{T}(X_1, X_2, X_3, Y)$.

To avoid additional technical description of n-ary concept-forming operators, we introduce n-adic formal concepts without their usage. The n-adic concepts of an n-adic context $(X_1, \ldots, X_n, Y)$ are exactly the maximal n-tuples $(A_1, \ldots, A_n)$ in $2^{X_1} \times \cdots \times 2^{X_n}$ with $A_1 \times \cdots \times A_n \subseteq Y$ with respect to component-wise set inclusion [85]. The notion of n-adic concept lattice can be introduced in the similar way to the triadic case [85]. For mining n-adic formal concepts one can use DATA-PEELER algorithm described in [12].

3.2 Biclustering

An alternative approach to define patterns in formal contexts can be realised via a relaxation of the definition of formal concept as a maximal rectangle full of crosses w.r.t. the input incidence relation. One of such relaxations is the notion of an object-attribute bicluster [37]. If $(g, m) \in I$, then (m', g') is called an *object-attribute bicluster*[2] (OA-bicluster or simply bicluster if there is no collision) with the density $\rho(m', g') = |I \cap (m' \times g')| / (|m'| \cdot |g'|)$.

The main features of OA-biclusters are listed below:

1. For any bicluster $(m', g') \subseteq 2^G \times 2^M$ it follows that $\frac{|m'| + |g'| - 1}{|g'||m'|} \leq \rho(A, B) \leq 1$.
2. OA-bicluster (m', g') is a formal concept iff $\rho = 1$.
3. If (m', g') is a bicluster, then $(g'', g') \leq (m', m'')$.

Let $(A, B) \subseteq 2^G \times 2^M$ be a bicluster and $\rho_{\min}$ be a non-negative real number such that $0 \leq \rho_{\min} \leq 1$, then (A, B) is called *dense*, if it fits the constraint $\rho(A, B) \geq \rho_{\min}$. The above-mentioned properties show that OA-biclusters differ from formal concepts by the fact that they do not necessarily have unit density. Graphically it means that not all the cells of a bicluster must be filled by crosses (see Fig. 2). The rectangle in Fig. 2 depicts a bicluster extracted from an object-attribute table. The horizontal grey line corresponds to object g and contains only non-empty cells. The vertical grey line corresponds to attribute m and also contains only non-empty cells. By applying the Galois operator, as explained in Sect. 2.1, one time to g we obtain all its attributes g'. By applying Galois operator $'$ twice to g we obtain all objects that

---

[2]We omit curly brackets here it what follows implying that $\{g\}' = g'$ and $\{m\}' = m'$.

Fig. 2 OA-bicluster

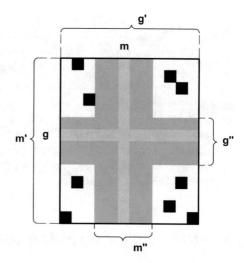

Algorithm 1: Add procedure for the online algorithm for OA-biclustering

---

Input: I is an input set of object-attribute pairs;
$\mathcal{B} = \{B = (*X, *Y)\}$ is a current set of OA-biclusters;
PrimesOA, PrimesAO;
Output: $\mathcal{B} = \{T = (*X, *Y)\}$;
PrimesOA, PrimesAO;
1: **for all** $(g, m) \in I$ **do**
2:      *PrimesOA*$[g]$:= *PrimesOA*$[g]$ $\cup$ m
3:      *PrimesAO*$[m]$:= *PrimesAO*$[m]$ $\cup$ g
4:      $\mathcal{B}$:= $\mathcal{B}$ $\cup$ (&*PrimesAO*$[m]$, &*PrimesOA*$[g]$)
5: **end for**

---

have the same attributes as g. This is depicted in Fig. 2 as g''. By applying Galois operator ' twice to m we obtain all attributes that belong to the same objects as m. This is depicted in Fig. 2 as m''. The white spaces indicate empty cells. The filled black boxes indicate non-empty cells. Whereas a traditional formal concept would cover only the green and grey area, the bicluster also covers the white and black cells. This gives to OA-biclusters fault-tolerance properties (see Proposition 1).

To generate biclusters fulfilling a minimal density requirement we can perform computations in two phases. The online phase, Add procedure (see Algorithm 1), allows to process pairs from incidence relation I and generate biclusters in one pass by means of pointer and reference variables for access to primes of objects and attributes even without knowing the number of objects and attributes in advance; see the version of this online algorithm for triadic case in [29]. Thus, the generation of all biclusters is realised within $O(|I|)$. Note that the algorithm can start with a non-empty collection of biclusters obtained previously. Then all biclusters can be enumerated in a sequential manner and only those fulfilling the minimal density constraint are retained.

For the context shown in Fig. 1 one can find two concepts,
$C_2 = (\{Alex, David\}, \{The\ Puppet\ Masters, Ubik\})$ and
$C_4 = (\{Alex, Mike, David\}, \{Ubik\})$, and one bicluster,
$B_1 = (Ubik', David') = (\{Alex, Mike, David\}, \{The\ Puppet\ Masters, Ubik\})$,
with density $\rho = 5/6 \approx 0.83$.

These two concepts can be interpreted as Sci-Fi readers and cyber punk readers (or P.K. Dick's readers at least), respectively. However, bicluster B_1 by allowing one missing pair (*Mike, The Puppet Masters*) can be considered as a community of Sci-Fi readers as well, which is larger than C_2.

3.3   OAC-Triclustering and Prime-Based n-Clustering

Guided by the idea of finding scalable and noise-tolerant alternatives to triconcepts, we have had a look at triclustering paradigm in general for a triadic binary data, i.e. for tricontexts as input datasets.

Definition 1 Suppose $\mathbb{K} = (G, M, B, I)$ is a triadic context and $Z \subseteq G, Y \subseteq M, Z \subseteq B$. A triple $T = (X, Y, Z)$ is called an *OAC-tricluster*. Traditionally, its components are called *extent, intent, and modus*, respectively.

The *density* of a tricluster $T = (X, Y, Z)$ is defined as the fraction of all triples of I in $X \times Y \times Z$:

$$\rho(T) = \frac{|I \cap (X \times Y \times Z)|}{|X||Y||Z|} \qquad (2)$$

Definition 2 A tricluster T is called *dense* iff its density is not less than some predefined threshold, i.e. $\rho(T) \geq \rho_{min}$.

The collection of all triclusters for a given tricontext $\mathbb{K}$ is denoted by $\mathscr{T}$.

Since we deal with all possible cuboids in Cartesian product $G \times M \times B$, it is evident that the number of all OAC-triclusters, $|\mathscr{T}|$, is equal to $2^{|G|+|M|+|B|}$. However, not all of them are supposed to be dense, especially for real data which are frequently quite sparse. Below we discuss one of the possible OAC-tricluster definitions, which gives us an efficient way to find, within polynomial time, a number of (dense) triclusters not greater than the number of triples in the initial data, $|I|$.

Here, let us define the prime operators and describe *prime OAC-triclustering*, which extends the biclustering method from [39] to the triadic case.

Derivation (prime) operators for elements of a triple $(\widetilde{g}, \widetilde{m}, \widetilde{b}) \in I$ from a triadic context $\mathbb{K}$ can be defined as follows:

$$\widetilde{g}' := \{ (m, b) \mid (\widetilde{g}, m, b) \in I \} \tag{3}$$

$$\widetilde{m}' := \{ (g, b) \mid (g, \widetilde{m}, b) \in I \} \tag{4}$$

$$\widetilde{b}' := \{ (g, m) \mid (g, m, \widetilde{b}) \in I \} \tag{5}$$

$(\widetilde{g}, \widetilde{m})'$, $(\widetilde{g}, \widetilde{b})'$, $(\widetilde{m}, \widetilde{b})'$ prime operators can be defined in the same way.

$$(\widetilde{g}, \widetilde{m})' := \{ b \mid (\widetilde{g}, \widetilde{m}, b) \in I \} \tag{6}$$

$$(\widetilde{g}, \widetilde{b})' := \{ m \mid (\widetilde{g}, m, \widetilde{b}) \in I \} \tag{7}$$

$$(\widetilde{m}, \widetilde{b})' := \{ g \mid (g, \widetilde{m}, \widetilde{b}) \in I \} \tag{8}$$

The following definition uses only prime operators [Eqs. (6)–(8)] to generate triclusters, however, other variants are possible. Thus, in [45], *OAC-triclusters based on box operator* have been studied; this type of tricluster relies on Eqs. (3)–(5).

Definition 3 Suppose $\mathbb{K} = (G, M, B, I)$ is a triadic context. For a triple $(g, m, b) \in I$ a triple $T = ((m, b)', (g, b)', (g, m)')$ is called a *prime operator based OAC-tricluster*. Its components are called *extent, intent, and modus*, respectively.

Prime-based OAC-triclusters are more dense than the ones based on box operator. Their structure is illustrated in Fig. 3: every element corresponding to the "grey" cell is an element of I. Thus, prime operator based OAC-triclusters in a three-dimensional matrix (tensor) form contain an absolutely dense cross-like structure of crosses (or ones).

The proposed OAC-tricluster definition has a fruitful property (see Proposition 1): for every triconcept in a given tricontext there exists a tricluster of the same

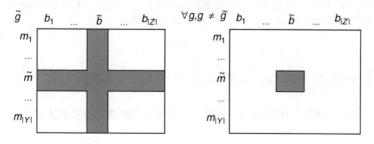

Fig. 3 Prime operator based tricluster structure

tricontext in which the triconcept is contained w.r.t. component-wise inclusion. It means that there is no information loss, we keep all the triconcepts in the resulting tricluster collection.

Proposition 1 Let $\mathbb{K} = (G, M, B, I)$ be a triadic context and $\rho_{\min} = 0$. For every $T_c = (X_c, Y_c, Z_c) \in \mathfrak{T}(G, M, B, I)$ with non-empty X_c, Y_c, and Z_c there exists a prime OAC-tricluster $T = (X, Y, Z) \in \mathscr{T}'(G, M, B, Y)$ such that $X_c \subseteq X, Y_c \subseteq Y, Z_c \subseteq Z$. Here, $\mathscr{T}'(G, M, B, I)$ denotes the set of all OAC-prime triclusters fulfilling the chosen value of $\rho_{\min}$.

Proof Let $(g, m, b) \in X_c \times Y_c \times Z_c$. By the definition of prime operators $(m, b)' := \{\widetilde{g} \mid (\widetilde{g}, m, b) \in I\}$. Since $m \in Y_c$ and $b \in Z_c$ then by the definition of formal triconcept (m, b) is related by Y to every $\widetilde{g} \in X_c$, therefore $(m, b)' \cap X_c = X_c$. Consequently for all $g_i \in X_c$ we have $g_i \in (m, b)'$. For $(g, b)'$ and $(g, m)'$ tricluster components the proof is similar. Finally, we have $X_c \subseteq X = (m, b)', Y_c \subseteq Y = (g, b)'$, and $Z_c \subseteq Z = (g, m)'$.

Prime-based n-clustering can be introduced similarly. Let $\mathbb{K} = (X_1, X_2, \ldots, X_n, Y)$ be an n-adic context and Y is binary relation between $X_1 \ldots X_n$.

Then for a tuple $(x_1, x_2, \ldots, x_n) \in Y$ we define n prime operators for each tuple $(x_1, \ldots, x_{i-1}, x_{i+1}, \ldots, x_n)$ as follows:

$$(\{x_1\}, \ldots, \{x_{i-1}\}, x_{i+1}, \ldots, \{x_n\})' = \{z_i \mid (x_1, \ldots, x_{i-1}, z_i, x_{i+1}, \ldots, x_n) \in Y\}.$$

For a given tuple $(x_1, x_2, \ldots, x_n) \in Y$, a prime operator based n-cluster is defined as follows:

$$P = ((\{x_2\}, \ldots, \{x_n\})', \ldots, (\{x_1\}, \ldots, \{x_{i-1}\}, \{x_{i+1}\}, \ldots, \{x_n\})', \ldots,$$

$$(\{x_1\}, \ldots, \{x_{n-1}\})').$$

The density of n-cluster $P = (Z_1, Z_2, \ldots, Z_n)$ is $\rho(P) = \frac{|Y \cap Z_1 \times Z_2 \times \ldots \times Z_n|}{|Z_1 \times Z_2 \times \ldots \times Z_n|}$. To keep analogy of ρ with physical density we refer to its numerator as the mass of P, i.e. mass(P), while its denominator plays a role of the volume of P, i.e. vol(P).

The description of a one-pass algorithm for OAC-prime tricluster generation can be found in [29]. A Map-Reduce based prototype of OAC-prime triclustering and possible implementation variants are presented in [94].

4   Quality Measures for Multimodal Clustering

4.1   Connection Between ρ and Local Clustering Coefficient

Since we use density as a local measure of n-cluster quality, it is useful to find its connection to local clustering coefficients (we use $cc_\bullet(\cdot)$ notation from [57]). For

$(V, E \subseteq V \times V)$, the local clustering coefficient is $cc_\bullet(v) = \frac{|N(v) \times N(v) \cap E|}{N(v)(N(v)-1)/2}$, here $N(v)$ is the degree of $v \in V$.

If one considers a one-mode network $(V, E \subseteq V \times V)$ as a formal context $\mathbb{K} = (G, G, I \subseteq G \times G)$, where $V = G$, and for $g, m \in V$ $gEm \iff gIm$, then for bicluster (g', g') it follows that[3]

$$\rho(g', g') = \frac{|g' \times g' \cap I|}{|g'||g'|} = \frac{|N(g) \times N(g) \cap I|}{|N^2(g)|} = \frac{|N(g) \times N(g) \cap I|}{\frac{(|N(g)|-1)|N(g)|}{2}} \frac{1 - 1/|N(g)|}{2}$$

$$= cc_\bullet(g) \frac{1 - \frac{1}{|N(g)|}}{2}.$$

Note that $N(g) = \deg(g) = \{u | gEu\} = g'$.

Moreover, for large neighbourhoods $\rho(g', g') \approx \frac{cc_\bullet(g)}{2}$.

4.2 Connection Between ρ and Modularity

Since we do not optimise any modularity-like criterion in our study, multimodal clusters are supposed to be overlapping in general, and, moreover, to the best of our knowledge there is no widely accepted modularity criterion even for bipartite overlapping communities; the introduction and study of such criteria could be a subject of a separate research. However, we show the interconnection between average of values in the input modularity matrix for a particular bicluster and its density.

Let A_{gm} be the adjacency matrix of an input context $\mathbb{K} = (G, M, I \subseteq G \times M)$, i.e. $A_{gm} = [gIm]$[4] for $(g, m) \in G \times M$. For bipartite graphs an entry of modularity matrix is defined as follows:

$$B_{gm} = A_{gm} - \frac{\deg(g)\deg(m)}{|I|} = [gIm] - \frac{|g'||m'|}{|I|}.$$

For non-overlapping communities modularity in two-mode networks is defined as follows [4]:

$$Mod = \frac{1}{|I|} \sum_{(g,m) \in G \times M} \left([gIm] - \frac{|g'||m'|}{|I|} \right) [(g, m) \in C], \text{ where}$$

---

[3]Note that technically (g', g') is not an OA-bicluster since $(g, g) \notin I$.

[4]Here $[\cdot]$ means Iverson bracket defined as $[P] = \begin{cases} 1 & \text{if } P \text{ is true;} \\ 0 & \text{otherwise,} \end{cases}$

$C \subseteq G \times M$ is a module (or community) from a set of non-overlapping communities $\mathscr{C}$ of the original network. Non-overlapping here is formally defined as follows: $\forall C, D \in \mathscr{C}$ $C \cap D = \emptyset$.

Let (m', g') be a bicluster of $\mathbb{K}$, then the sum over all entries $(\widetilde{g}, \widetilde{m}) \in m' \times g'$ in B gives

$$|m' \times g' \cap I| - \frac{\sum\limits_{(\widetilde{g}, \widetilde{m}) \in m' \times g'} |\widetilde{g'}||\widetilde{m'}|}{|I|}.$$

Instead of normalising that sum by $|I|$ as in modularity definition, we can try to calculate (local) bicluster modularity, $\mathrm{Mod}_l(m', g')$, by normalising the sum by the bicluster volume $\mathrm{Vol}(m', g') = |g'||m'|$:

$$\mathrm{Mod}_l(m', g') = \frac{|m' \times g' \cap I|}{|g'||m'|} - \frac{\sum\limits_{\widetilde{g} \in m'} |\widetilde{g'}| \sum\limits_{\widetilde{m} \in g'} |\widetilde{m'}|}{|g'||m'||I|} = \rho(m', g') - \frac{\overline{\deg(\widetilde{g})}\,\overline{\deg(\widetilde{m})}}{|I|}, \text{ where}$$

$\overline{\deg(\widetilde{g})} = \frac{\sum_{\widetilde{g} \in m'}}{|g'|}$ is the average degree of $\widetilde{g}$ in the input bicluster and $\overline{\deg(\widetilde{m})}$ is the average degree of $\widetilde{m}$ and defined similarly.

It is clear that to maximise Mod_l criterion one needs to find a bicluster with high density and low average degrees of its elements.

However, the original modularity criterion for bipartite non-overlapping networks has intrinsic drawbacks. The first problem, low resolution, consists in the dependence between the size of detected communities and the size of an input graph [21]. Another one can be demonstrated by a model example.

Let $\mathbb{K} = (G, M, I)$ be a formal context, where for a certain pair $(g, m) \in I$ we have $g' = M, m' = G$, and $I = m' \times m \cup g \times g'$. Without loss of generality let $|G| = |M| = n$. Then

$$B_{gm} = [gIm] - \frac{|g'||m'|}{|I|} = 1 - \frac{n^2}{2n - 1}.$$

For large n, $B_{gm} \approx 1 - n/2$ and this value tends to $-\infty$ by implying $n \to \infty$. To keep the second term of an entry of the modularity matrix no greater than 1 (the maximal probability of incidence of g and m), one needs to require $|g'|, |m'| \leq \sqrt{|I|}$ (which is in fact should be normally fulfilled for large and sparse (real) networks).

4.3 Least Square Optimal n-Clusters

One of the important statistics in clustering is the data scatter of an input matrix, i.e. the sum of squares of all its entries [63]. In [64], least squares based maximisation criterion to generate n-cluster was proposed:

$$g(P) = \rho^2(P) \cdot \text{Vol}(P) = \rho(P) \cdot \text{mass}(P), \text{ where}$$

P is an n-cluster of a certain n-adic context. On the one hand, its direct interpretation implies that we care about dense n-clusters of large size instead of only dense (that may be small) or only large (that may be sparse); in other words such n-clusters tend to be massive (with low number of missing tuples in the input binary relation) and dense. On the other hand, this criterion measures the contribution of P to the data scatter of the input n-adic context.

In [45], one can find a theorem saying that by maximisation of $g(P)$ we require higher density within n cluster P than in the corresponding outside regions along its dimensions.

4.4 Weak Bicluster Communities and Graph Cuts

In network analysis, a community is called weak if its average internal degree is greater than its average out degree [3].

In two-mode case, for an input context $\mathbb{K} = (G, M, I)$ and its bicluster (m', g'), we have

$$\sum_{\widetilde{g} \in m'} |(\{\widetilde{g}\} \cup \{g\})'| + \sum_{\widetilde{m} \in g'} |(\{\widetilde{m}\} \cup \{m\})'| \geq \sum_{\widetilde{g} \in m'} |\widetilde{g'} \cap M \setminus g'| + \sum_{\widetilde{m} \in g'} |\widetilde{m'} \cap G \setminus m'|.$$

The left-hand side of the inequality is the doubled sum of the number of object-attribute pairs from (m', g'). The right-hand side shows how many pairs the objects from bicluster extent and the attributes from bicluster intent form with the remaining attributes and objects of the context, respectively. In network analysis this measure is known as *cut* [21], i.e. the number of edges one should delete to make the community disconnected from the remaining vertices in the input graph. Thus, the inequality can be rewritten as follows:

$$\rho(m', g') \geq \frac{\text{cut}(m', g')}{2|g'||m'|}.$$

This criterion can be used for selection of biclusters during their generation instead of fixed ρ_{min}.

4.5   Stability of OA-Biclusters

Stability of formal concepts [53, 54] has been used as a means of concepts' filtering in studies on epistemic communities [56, 77, 78] and communities of website visitors [55].

Let $\mathbb{K} = (G, M, I)$ be a formal context and (A, B) be a formal concept of $\mathbb{K}$. The *(intensional) stability index*, σ, of (A, B) is defined as follows:

$$\sigma(A, B) = \frac{|\{C \subseteq A \mid C' = B\}|}{2^{|A|}}$$

As we know, not all of the OA-biclusters of a given formal context are formal concepts.

Only those OA-biclusters that fulfill condition $(m', g') = (g'', m'')$ are formal concepts. However, stability index can be technically computed for any OA-bicluster as follows:

$$\sigma(m', g') = \frac{|\{A \subseteq m' \mid A' = g'\}|}{2^{|m'|}}$$

Set $2^{m'}$ can be decomposed into three parts: $2^{g''} \cup 2^{m' \setminus g''} \cup \Delta$. The numerator is equal to $|\{A \in 2^{g''} \mid A' = g'\}| + |\{A \in 2^{m' \setminus g''} \mid A' = g'\} \setminus \emptyset| + |\{A \in \Delta \mid A' = g'\} \setminus \emptyset|$. Since every set of objects from $m' \setminus g''$ does not have all attributes from g', the second summand is 0, and the same applies to the third one due to each set from Δ contains at least one object $\widetilde{g}$ from $m' \setminus g''$ such that $\widetilde{g}' \neq g'$. Hence,

$$\sigma(m', g') = \frac{|\{A \in 2^{g''} \mid A' = g'\}|}{2^{|m'|}}.$$

Since the number of all A that contain g is $|2^{g'' \setminus g}|$, the tight lower bound of OA-bicluster's stability is $2^{|g'' \setminus g| - |m'|}$.

The stability index of a concept indicates how much the concept intent depends on particular objects of the extent.

4.6   Coverage and Diversity

Diversity is an important measure in Information Retrieval for diversified search results and in Machine Learning for ensemble construction [82].

To define diversity for multimodal clusters we use a binary function that equals to 1 if the intersection of triclusters T_i and T_j is not empty, and 0 otherwise.

$$\text{intersect}(T_i, T_j) = \left[G_{T_i} \cap G_{T_j} \neq \emptyset \wedge M_{T_i} \cap M_{T_j} \neq \emptyset \wedge B_{T_i} \cap B_{T_j} \neq \emptyset \right] \qquad (9)$$

It is also possible to define *intersect* for the sets of objects, attributes, and conditions. For instance, $\text{intersect}_G(T_i, T_j)$ is equal to 1 if triclusters T_i and T_j have non-empty intersection of their extents, and 0 otherwise.

Now we can define *diversity of the tricluster set* $\mathscr{T}$:

$$\text{diversity}(\mathscr{T}) = 1 - \frac{\sum_j \sum_{i<j} \text{intersect}(T_i, T_j)}{\frac{|\mathscr{T}|(|\mathscr{T}|-1)}{2}} \tag{10}$$

The *diversity for the sets of objects (attributes or conditions)* is similarly defined:

$$\text{diversity}_G(T) = 1 - \frac{\sum_j \sum_{i<j} \text{intersect}_G(T_i, T_j)}{\frac{|\mathscr{T}|(|\mathscr{T}|-1)}{2}} \tag{11}$$

Coverage is defined as a fraction of the triples of the context (alternatively, objects, attributes, or conditions) included in at least one of the triclusters of the resulting set.

More formally, let $\mathbb{K} = (G, M, B, I)$ be a tricontext and $\mathscr{T}$ be the associated triclustering set obtained by some triclustering method, then coverage of $\mathscr{T}$:

$$\text{coverage}(\mathscr{T}) = \sum_{(g,m,b)\in I} \left[(g, m, b) \in \bigcup_{(X,Y,Z)\in\mathscr{T}} X \times Y \times Z \right] / |I|. \tag{12}$$

The *coverage of the object set* G by the tricluster collection $\mathscr{T}$ is defined as follows:

$$\text{coverage}_G(\mathscr{T}) = \sum_{g\in G} \left[g \in \bigcup_{(X,Y,Z)\in\mathscr{T}} X \right] / |G|. \tag{13}$$

Coverage of attribute or condition sets can be defined analogously. These measures may have sense when one would like to know how many actors or items in the network do not belong to any found community.

We also use the *coverage of formal concepts by biclusters*, i.e. we count the number of concepts covered by at least one bicluster in the corresponding bicluster collection B. We say that bicluster $B = (X, Y)$ covers concept $C = (Z, W)$ w.r.t. component-wise inclusion of their extents and intents, namely $C \sqsubseteq B : \iff Z \subseteq X$ and $W \subseteq Y$.

$$\text{coverage}_{\mathscr{B}}(\mathfrak{B}(G, M, I)) = \frac{\{C \in \mathfrak{B}(G, M, I) \mid \exists B \in \mathscr{B} : C \sqsubseteq B\}}{|\mathfrak{B}(G, M, I)|}. \tag{14}$$

5   Data

For our experiments we collected datasets from one-mode to four-mode networks.
   In particular, we have analysed the following classic one-mode datasets:

- Karate club, 34×34, 78 edges;
- Florentine family 1, 16×16, 40 edges;
- Florentine family 2, 16×16, 30 edges;
- Hi-tech, 36×36, 147 edges;
- Mexican people, 35×35, 117 edges.

For two-mode datasets we have used Southern women of size 18×14 with 93
edges and four datasets studied in [57]:

- co-authoring, 19,885×16,400, and 45,904 edges;
- co-occurrence, 13,587×9,263, and 1,833,63 edges;
- actor, 127,823×383,640, and 1,470,418 edges;
- p2p, 1,986,588 peers×5,380,546 data, and 55,829,392 links (edges).

As for three-mode network, we have analysed Bibsonomy dataset[5] with $|U| =$
2467 users, $|T| = 69,904$ tags, $|R| = 268,692$ resources that related by $|Y| =$
816,197 triples.
   Finally, MovieLens data[6] with 100,000 ratings (integers from 1 to 5) and 1300
tag applications applied to 9000 movies by 700 users is considered as a four-mode
dataset. We have used only user, movie, rating, and time modes.

6   Experiments

We have tested our implementations for one- and two-mode networks in Python 2.7
and for higher modes in C# with our tool, Multimodal Clustering Toolbox, on a Mac
Pro computer with 3.7 GHz and 16 GB RAM.

6.1   Two-Mode Networks

For each two-mode dataset we report the number of unique biclusters and the
number of all generated biclusters; note that when all objects (and attributes) are
pairwise different there are no duplicates by definition.

---

[5]http://www.kde.cs.uni-kassel.de/bibsonomy/dumps/.
[6]http://grouplens.org/datasets/movielens/.

Table 1 Southern women: 18×14, 93 edges

ρ	Concept coverage	Unique biclusters	Biclusters	Fraction of covered concepts
0	65	83	93	1.00
0.05	65	83	93	1.00
0.1	65	83	93	1.00
0.15	65	83	93	1.00
0.2	65	83	93	1.00
0.25	65	83	93	1.00
0.3	65	83	93	1.00
0.35	65	82	92	1.00
0.4	65	81	91	1.00
0.45	65	77	87	1.00
0.5	65	71	81	1.00
0.55	65	63	73	1.00
0.6	65	60	7	1.00
0.65	64	51	59	0.98
0.7	63	40	47	0.97
0.75	57	33	4	0.88
0.8	51	22	28	0.78
0.85	35	13	19	0.54
0.9	20	7	9	0.31
0.95	0	0	0	0.00
1	0	0	0	0.00

For small and medium size classic two-mode and one-mode datasets we have reported the number of formal concepts covered by the generation bicluster collection for a specific ρ_{min} as well as their fraction, i.e. coverage$_{\mathscr{B}}(\mathfrak{B}(G, M, I))$.

In 1930s, a group of ethnographers collected data on the social activities of 18 women over a nine-month period [17]. Different subgroups of these women had met in 14 informal social events; the incidence of a woman to a particular event was established using "interviews, the records of participant observers, guest lists, and the newspapers" [17, p. 149]. Later on, this Southern Women dataset has become a benchmark for comparing communities detection methods in two-mode social network analysis, in particular, including concept lattices as a community detection approach [22, 23].[7]

The results of our experiments with Southern Women dataset are summarised in Table 1.

---

[7]There is a small inconsistency in the profiles of women w_{14} (Helen) and w_{15} (Dorothy), namely between their description in [22] and the downloaded dataset provided at https://networkdata.ics. uci.edu/netdata/html/davis.html, thus according to the latter $e_{12}, e_{13} \in w'_{14}$ and $e_{11}, e_9 \in w'_{15}$.

There are 66 formal concepts for the Southern woman network. Since OA-biclusters are tolerant to missing values, let us illustrate how rather dense biclusters include the largest concepts with non-empty extent and intent.

For example, with $\rho_{min} = 0.8$ we show five bicluster-concept pairs $B_i = (e', w')$, $C_i = (W, E)$ related by component-wise inclusion of their extents and intents, respectively, namely $C_i \sqsubseteq B_i : \iff W \subseteq e'$ and $E \subseteq w'$:

1. $C_1 = (\{w_0, w_1, w_2, w_3, w_5, w_6, w_7\}, \{e_5, e_7\}) \sqsubseteq B_1 = (\{w_0, w_1, w_2, w_3, w_5, w_6, w_7, w_8\}, \{e_2, e_4, e_5, e_7\})$ with $\rho(B_1) = 0.84$;
2. $C_2 = (\{w_0, w_2, w_3\}, \{e_2, e_3, e_4, e_5, e_7\}) \sqsubseteq B_2 = (\{w_0, w_2, w_3, w_4\}, \{e_0, e_2, e_3, e_4, e_5, e_6, e_7\})$ with $\rho(B_2) = 0.82$;
3. $C_3 = (\{w_9, w_{10}, w_{11}, w_{12}, w_{13}, w_{14}, w_{15}\}, \{e_{11}\}) \sqsubseteq B_3 = (\{w_9, w_{10}, w_{11}, w_{12}, w_{13}, w_{14}, w_{15}\}, \{e_6, e_7, e_8, e_{11}\})$ with $\rho(B_3) = 0.82$;
4. $C_4 = (\{w_{10}, w_{11}, w_{12}, w_{15}\}, \{e_7, e_8, e_9, e_{11}\}) \sqsubseteq B_4 = (\{w_{10}, w_{11}, w_{12}, w_{13}, w_{14}, w_{15}\}, \{e_7, e_8, e_9, e_{11}\})$ with $\rho(B_4) = 0.92$;
5. $C_5 = (\{w_{16}, w_{17}, w_{13}\}, \{e_1, e_8\}) \sqsubseteq B_5 = (\{w_{16}, w_{17}, w_{13}, w_{14}\}, \{e_1, e_8\})$ with $\rho(B_5) = 0.88$.

The corresponding bipartite graph is shown in Fig. 4 along with the biclique formed by elements of concept C_1 and bicluster B_1, and concept C_3 and bicluster B_3. According to [18, 22] there is the "true structure" of the Southern women network: namely, there are two groups of women $\{w_0, \ldots, w_8\}$ and $\{w_1, \ldots, w_{17}\}$. The first group of women participated in events e_0 through e_4, while the second group was not. The second group participated in events e_3 through e_{13}, while the first group was not. Both groups participated e_6, e_7, and e_8.

Since the Southern women network is a well-studied case in SNA community and one of the first SNA datasets analysed by sociologists using concept lattices, an interested reader may refer to [22, 23] to find professional interpretation of several important communities of women found by means of formal concepts.

Even though that such networks as co-authoring, co-occurrence, actor, and p2p are two-mode and known to SNA community about a decade, even the number of concepts (maximal bicliques) for these datasets is not reported in the literature (Tables 2 and 3).

An interesting issue has appeared: At which ρ_{min} the generated biclusters do not cover all formal concepts with non-empty extent and intent? According to our experiments for two-mode (see also Appendix) and one-mode networks, it usually happens around $\rho_{min} = 0.5$ or higher (containing intervals marked by two horizontal lines in the tables), so, we may hypothesise that one can normally set minimal density value equal to 0.5.

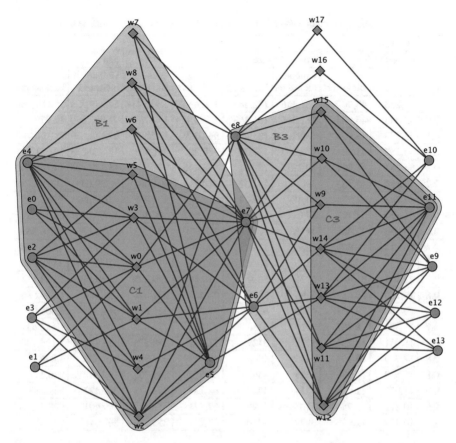

Fig. 4 The two-mode network for the Southern women dataset, bicluster B_1 and concept C_1, and bicluster B_3 and concept C_3

6.2 Folksonomies as Three-Mode Networks

Folksonomy is a typical example of a three-mode network, where a hyperedge connects a user, a tag, and an attribute. Thus each hyperedge is a set of size three with three vertices of different types; it is convenient to represent edges as tuples (*user, tag, resource*). Since we experiment with Bibsonomy, a Folksonomy-based resource sharing system for scientific bibliography, our users are scientists, resources are papers that they bookmarked or even authored; a tag is assigned by a scientist to a particular paper while bookmarking.

Table 2 The numbers of unique and all OA-biclusters for the four large two-mode networks

| | Datasets | | | | | | | |
| | Co-authoring | | Co-occurrence | | Actor | | p2p | |
ρ	Unique biclusters	Biclusters	Unique biclusters	Biclusters	Unique biclusters	Biclusters	Unique biclusters	Biclusters
0	43,253	45,904	161,386	183,363	1,278,989	1,470,418	54,789,256	55,829,169
0.05	43,253	45,904	161,386	183,363	1,226,429	1,417,827	41,937,580	42,973,016
0.1	43,253	45,904	160,200	181,630	962,389	1,153,704	27,178,639	28,196,480
0.15	43,253	45,904	124,383	137,367	700,207	891,401	18,320,253	19,321,315
0.2	43,251	45,902	69,283	75,761	523,446	714,509	13,179,196	14,165,402
0.25	43,184	45,835	39,081	43,252	410,118	601,065	9,789,039	10,759,880
0.3	42,748	41,774	24,484	27,672	318,245	509,068	7,019,097	7,969,965
0.35	41,774	44,423	17,011	19,718	269,642	460,361	5,088,606	6,017,582
0.4	39,366	42,008	12,796	15,100	214,979	405,543	3,950,659	4,856,567
0.45	36,194	38,809	10,111	12,251	190,704	381,106	3,369,522	4,261,678
0.5	34,141	36,737	8539	10,515	182,906	373,191	3,056,597	3,938,536
0.55	29,404	31,960	6926	8699	110,464	299,895	1,156,887	1,918,111
0.6	23,150	25,615	5395	7036	84,459	272,894	764,584	1,483,586
0.65	20,604	23,007	4572	6127	77,904	265,699	614,743	1,308,939
0.7	16,391	18,707	3929	5386	72,651	259,877	50,981	1,182,631
0.75	15,951	18,234	3726	5129	71,663	258,550	472,869	1,126,702
0.8	12,989	15,137	3490	4846	69,449	255,904	419,533	1,046,786
0.85	11,533	13,530	3313	4568	68,555	254,703	39,189	986,811
0.9	11,053	12,976	3214	4437	68,186	254,138	377,377	949,637
0.95	10,875	12,756	3105	4290	67,871	253,623	369,401	929,765
1	10,874	12,756	3079	4250	67,798	253,390	367,946	926,380

Table 3 Elapsed time for online OA-biclustering

Dataset	$|I|$	$G|$	$|M|$	Time, s
Co-authoring	45,904	19,885	16,400	0.13
Co-occurrence	183,363	13,587	9264	0.25
Actor	1,470,418	127,823	383,640	3.55
p2p	55,829,392	19,86,588	5,380,546	260.13

Let us consider a toy imaginary example of Bibsonomy data; the input context is shown by three layers in Table 4. There are four users (u_1 = *Fortunato*, u_2 = *Freeman*, u_3 = *Newman*, and u_4 = *Roth*) and three tags (t_1 = *Galois Lattices*, t_2 = *SNA*, and t_3 = *Statistical Physics*). Three papers p_1, p_2, and p_3 are marked according to the research interests of those users. Thus Freeman and Roth marked paper 1 by tags "Galois Lattices" and "SNA",

(continued)

while Fortunato and Newnam tagged paper 3 by tags 'SNA" and "Statistical Physics". All the users assigned tag "SNA" to paper 2. Three corresponding communities can be easily captured by formal triconcepts:

$$C_1 = (\{u_2, u_4\}, \{t_1, t_2\}, \{p_1\})$$
$$C_2 = (\{u_1, u_3\}, \{t_2, t_3\}, \{p_3\})$$
$$C_3 = (\{u_1, u_2, u_3, u_4\}, \{t_2\}, \{p_2\}).$$

Concept C_3 is more general than C_1 and C_2 w.r.t. extent inclusion, and corresponds to SNA-interested users, while C_1 corresponds to those who are interested in concept lattices for SNA domain, and C_2 unites users interested in SNA by means of methods similar to their prototypes in Statistical Physics. The corresponding hypergraph with these triconcepts is shown in Fig. 5.

Table 4 A toy example with Bibsonomy data

	t_1	t_2	t_3
u_1			
u_2	×	×	
u_3			
u_4	×	×	
	p_1		

	t_1	t_2	t_3
u_1		×	
u_2		×	
u_3		×	
u_4		×	
	p_2		

	t_1	t_2	t_3
u_1		×	×
u_2			
u_3		×	×
u_4			
	p_3		

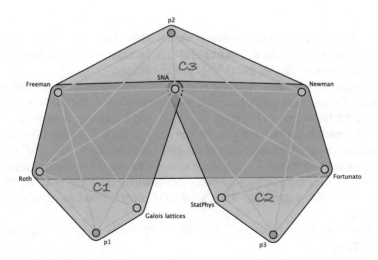

Fig. 5 Three triconcepts C_1, C_2, C_3 for the Bibsonomy three-mode network

Table 5 Experimental results for k first triples of Bibsonomy dataset with $\rho_{min} = 0$

k, number of first triples	$|U|$	$|T|$	$|R|$	$|\mathfrak{T}|$	$|\mathscr{T}_{OAC'}|$	TRIAS, s	OAC-Prime,s	
							Full time	Online phase
100	1	47	52	57	77	0.2	0.02	0.003
1000	1	248	482	368	656	1	0.043	0.001
10,000	1	444	5193	733	1461	2	273	0.031
100,000	59	5823	28,920	22,804	33,172	3386	24,185	0.542
200,000	340	14,982	61,568	–	105,571	> 24 h	25,446	1.268
500,000	1191	45,232	148,695	–	316,139	> 24 h	29,035	3.529
816,197	2467	69,904	268,692	–	484,349	> 24 h	241,341	5.186

Table 6 Density distribution of OAC-prime triclusters for 816,197 triples of Bibsonomy dataset with $\rho_{min} = 0$

Lower bound of ρ	Upper bound of ρ	Number of triclusters
0	0.05	172
0.05	0.1	3070
0.1	0.2	36,878
0.2	0.3	77,170
0.3	0,4	90,005
0.4	0.5	67,659
0.5	0.6	66,711
0.6	0.7	41,507
0.7	0.8	22,225
0.8	0.9	11,662
0.9	1	67,290

To build all triconcepts of a certain context we have used a Java implementation of the TRIAS algorithm by R. Jäschke [46]. The last two columns in Table 5 mean time of execution of TRIAS and OAC-prime algorithms.

Note that here we have reported both the full execution time of OAC-prime algorithm, i.e. tricluster generation with density calculation, and the time of online phase for tricluster generation only. One may note a dramatical drop-off in time efficiency between the last and penultimate lines in Table 5 for the full execution time, while online phase took only about half a second more. The devil is in the hashing data structures used for duplicate elimination and we believe the timing can be improved, for example, by a specially designed Bloom filter. Note that a more general and efficient algorithm Data-Peeler [13] could be used suitable for mining n-concepts.

Distribution of density of triclusters for all the triples of Bibsonomy dataset is given in Table 6.

6.3   MovieLens Data as Four-Mode Network

We summarise the results of prime-based tetraclustering execution on Movielens data below:

Time:	$13,252$ ms
Number of n-clusters:	89,931
Average volume, $\overline{Vol}$:	455.4
Average density, $\overline{\rho}$:	0.35
Average coverage:	0.1%
Average mass, $\overline{mass}$:	103.7
Average $\rho \cdot mass$:	28.1

In addition to average density we report average volume, average coverage (the number of covered original tuples by each tetracluster on average), average mass (the number of tuples inside each tetraclusters on average), and quite an interesting statistic, average $\rho \cdot mass$. If we maximise the latter criterion, then we require for our tetraclusters to be dense and large at the same time while criterion $\rho \cdot Vol$ could result in sparse patterns.

To provide concrete examples of tetraclusters, we have selected rather small-sized dense communities in Table 7.

Table 7 Tetraclusters for Movielens data

No.	Generating tuple	Volume	ρ	Coverage	$mass$	$\rho \cdot mass$
1	(483, Star Trek IV, 5, 1997/11)	27	0.93	0.03 %	25	23.1
2	(384, Evita, 5, 1998/03)	15	0.87	0.01 %	13	11.3
3	(872, Scream 2, 5, 1998/02)	15	0.87	0.01 %	13	11.3
4	(102, Face/Off, 3, 1997/10)	12	0.92	0.01 %	11	10.1
5	(750, Gang Related, 1, 1997/11)	9	1.00	0.01 %	9	9.0

No.	Users	Movies	Rating	Time
1	{109,307,374,483, 87,545,815,882,927}	{Star Trek: The Wrath of Khan (82), Star Trek IV: The Voyage Home (86), Star Wars (77) }	{5}	{97/11}
2	{378,384,392}	{Good Will Hunting (97), Evita (96), Titanic (97), L.A. Confidential (97), As Good As It Gets (97)}	{5}	{98/03}
3	{206,332,872}	{Time to Kill, A (96), Scream (96), Scream 2 (97), Air Force One (97), Titanic (97)}	{5}	{98/02}
4	{102,116,268,430}	{Grosse Pointe Blank (1997), Face/Off (1997) } Air Force One (1997)}	{3}	{97/10}
5	{181,451,750}	{Gang Related (1997), Rocket Man (1997) Leave It to Beaver (1997)}	{1}	{97/11}

For example, one can easily identify the community of modern space opera lovers in 4-cluster no. 1. Note that their third and fourth components are always sets containing a single element due to the chosen mode nature: the same people cannot rate the same movies by different marks simultaneously or within a different month.

6.4   One-Mode Networks as Two-Mode Ones

There are different techniques called projections to transform two-mode graphs to their one-mode versions [57, 67]. Sometimes, researchers even do transformations in backward direction to consider interactions between different subgroups of actors as they were from different modes of the corresponding two-mode network [18, 91].

An undirected one-mode network in the form $\Gamma = (G, E \subseteq G \times G)$ can be considered as the two-mode network by composing a context $\mathbb{K} = (G, G, I)$ where $gEh \iff gIh$ for any $g, h \in G$, with two options for I being a symmetric relation: a) reflexive and b) irreflexive.

In reflexive case, each concept (A, B) of such context $\mathbb{K}$ that fulfills $A = B$ corresponds to the maximal clique A in the original one-mode network.

We provide the reader with the results of OA-biclustering for one-mode networks in Tables 8, 9, 10, 11, and 12.

In addition to the fraction of covered concepts by component-wise set inclusion we have reported intervals $[\rho_\alpha, \rho_\beta]$, where the fraction of covered concepts decreases below 1 first time for each dataset (see two vertical lines in the tables).

In addition to the reported statistics, let us demonstrate found biclusters and concepts for Zachary's karate club dataset. Originally, the author of [90], an anthropologist, described social relationships between members of a karate club in the period of 1970–72; the network contains 34 active members of the karate club who interacted outside the club, including 78 pairwise links between them. The club was split into two parts after a conflict between its instructor and president. This dataset is usually used as a benchmark for demonstration and testing of community detection algorithms [3].

In Fig. 6, one can see three biclusters (B_1, B_2, and B_3) with density less than 1 but greater 0.8 each. Thus none of them is a concept; moreover, union of their intent and extent does not form a clique of the input one-mode network.

$$B_1 = (29', 29') = (\{32, 33, 26, 29, 23\}, \{32, 33, 26, 29, 23\}) \text{ with } \rho = 0.84$$

$$B_2 = (3', 12') = (\{0, 1, 2, 3, 7, 12, 13\}, \{0, 3, 12\}) \text{ with } \rho = 0.81$$

$$B_3 = (5', 4') = (\{0, 10, 4, 6\}, \{0, 10, 4, 5\}) \text{ with } \rho = 0.88$$

(continued)

Among all generated concepts, each concept (X, Y) with $X = Y$ results in clique X.

Thus concept $(\{0, 1, 2, 3, 7\}, \{0, 1, 2, 3, 7\})$ forms clique $Q_1 = \{0, 1, 2, 3, 7\}$, while concepts $(\{0, 1, 2, 3, 13\}, \{0, 1, 2, 3, 13\})$ and $(\{32, 33, 29, 23\}, \{32, 33, 29, 23\})$ result in $Q_2 = \{0, 1, 2, 3, 13\}$ and $Q_3 = \{32, 33, 29, 23\}$, respectively. Those are cliques of maximal size 5 and 4 from two parts of the karate club after its fission. It is evident that for each of those cliques its set of vertices can be found in some OA-bicluster. One can check that the set of vertices of B_1 contains those of Q_3, and vertices of B_2 include those of Q_1 and Q_2. So, it is possible to conclude that even though the density of a bicluster may be less than 1, they can contain more vertices resulting in larger communities than cliques. Note that the club instructor, 0, belongs to extents of B_2 and B_3 being a "missing link" between two corresponding subcommunities, which lack in active interaction otherwise.

Table 8 Karate club: 34×34, 190 edges

ρ	Covered concepts	Unique biclusters	Biclusters	Fraction of covered concepts
0	134	190	190	1.00
0.05	134	190	190	1.00
0.1	134	190	190	1.00
0.15	134	190	190	1.00
0.2	134	190	190	1.00
0.25	134	190	190	1.00
0.3	134	184	184	1.00
0.35	134	178	178	1.00
0.4	134	163	163	1.00
0.45	134	142	142	1.00
0.5	132	128	128	0.99
0.55	126	108	108	0.94
0.6	115	91	91	0.86
0.65	97	71	71	0.72
0.7	90	67	67	0.67
0.75	68	47	47	0.51
0.8	31	25	25	0.23
0.85	27	20	20	0.20
0.9	12	12	12	0.09
0.95	12	12	12	0.09
1	12	12	12	0.09

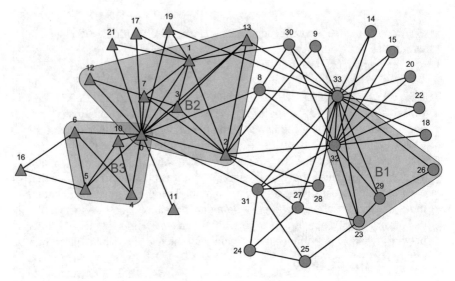

Fig. 6 Three dense biclusters B_1, B_2, B_2 found in Karate club network with $\rho_{min} = 0.8$

In Fig. 7, one can see three found communities that are composed of vertices corresponding to three concepts C_1, C_2, and C_3.

$$C_1 = (\{32, 33\}, \{32, 33, 8, 14, 15, 18, 20, 22, 23, 29, 30, 31\})$$

$$C_2 = (\{0, 1\}, \{0, 1, 2, 3, 7, 13, 17, 19, 21\})$$

$$C_3 = (\{0, 10, 6\}, \{0, 4, 5\})$$

In this concrete example, the usage of formal concepts for representing communities seems to be even more beneficial than that of dense OA-biclusters since we have been able to cover almost both parts of the separated karate club by three concepts without sharing members between the counterparts; concepts C_1 and C_2 contain more vertices than biclusters B_1 and B_2 shown in Fig 6. Note that the semantic of C_1 lies in the interpretation of its intent as common contacts of 32 and 33, an active club member who is loyal to the club's president and the president, respectively. Intent of C_2 contains members mutually connected with the club instructor, 0, and member 1.

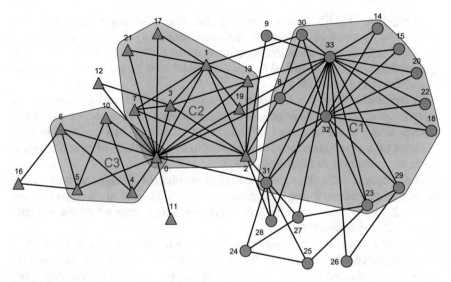

Fig. 7 Three formal concepts C_1, C_2, C_2 found in Karate club network

7 Related Work

There is a so-called subspace clustering [1] closely related to biclustering, where objects are considered as points in high dimensional space and clustered within multidimensional grid of a certain granularity. However, these methods cannot be directly applied to multidimensional relational data, i.e. multi-mode networks, since entities from different modes are often numbered arbitrarily and do not follow a pre-specified order like values along numerical axes. However, biclustering of numerical data, which may describe two-mode weighted networks, can be realised with Triadic Concept Analysis in case we consider attribute values as a mode of conditions under which an object has an attribute [50]. These results are also applicable to n-dimensional numerical datasets. Two other ways to deal with numeric data are to apply the so-called scaling, e.g., using a binary threshold, or Pattern Structures defined on vectors of numeric intervals [16, 25, 49]. Pattern Structures were also used to rethink collaborative filtering and find relevant taste communities for a particular user in terms of vectors of desirable rating intervals for good movies [34].

As for OA-biclustering, it has been used in several applications; for example, OA-biclustering has been applied for finding market segments in two-mode data on Internet advertising to recommend advertising terms to companies playing on these segments [35, 39]. In crowdsourcing platforms, OA-biclustering helps to find similar ideas (proposals) to discuss potential collaborators [43, 44] as well as answer questions [14]; in case we consider opinions of users over a set of different ideas (proposals), it is possible to find antagonists, which may be prospective opponents in crowdsourcing teams [41].

In fact, biclustering is a well-established tool in Bioinformatics, especially for Gene Expression Analysis in genes-samples networks [49, 70]. A non-exhaustive concept lattice based taxonomy of biclustering techniques can be found in [36]. Methods for three-mode networks are applicable in this domain when in addition to genes and samples time mode comes [92].

Going back to networks, several researchers define other kinds of networks where the role of dimensions is played by different types of labels of multi-edges between actors [8, 9]; they call such networks multidimensional while others use the term multi-relational networks [88].

One more variation of networks is realised by n-partite networks where connection are edges between vertices of allowed types [80]. It is possible to mine maximal closed and connected subgraphs in them and interpret them as communities [59]; these patterns coincide with bicliques and formal concepts in two-mode case. However, for higher dimensions such n-partite graphs are not equivalent to n-adic contexts and may result in information loss or phantom hyperedges if we reduce the latter to the former or vice versa [33]. In [28], for analysing such tripartite network composed by two two-mode networks with one shared part, biclusters from these two networks have been used. Namely, those biclusters that are similar with respect to their extents are merged by taking the intersection of their extents. The intent of the first bicluster and the intent of the second bicluster become the intent and modus, respectively, of the resulting tricluster. In FCA domain, analysis of n-partite and multi-relational networks can be unified within Relational Concept Analysis where objects can be involved in different types of relations with attributes and each other [30].

Another related subject is tensor factorisation, which is of high importance in Data Mining [71] and Machine Learning [15] due to its ability to reduce data dimensionality, find the so-called hidden factors, and even perform information fusion. The closest approaches to ones in the presented study can be found in works on Boolean matrix [6, 7] and tensor factorisation [5, 62]. Thus in [7] it was shown that formal concepts may result in optimal factors in Boolean matrix decomposition; in [2, 42] these decompositions showed their competitive applicability to collaborative filtering by finding communities of similar tastes. Tensor clustering is another way to find dense patterns; this approach is very similar to multimodal clustering in n-ary relations, especially in case of Boolean tensors, which normally represent n-ary relations between entities [38, 61, 64, 79]. An interesting issue here, whether it is possible to obtain improvements in classification accuracy for tensors with labeled objects from one of their dimensions over conventional object-attribute representations [93].

Since the proposed multimodal clustering is an approach to find approximate patterns, not absolutely dense as closed n-sets or n-adic concepts, various similar ideas can be proposed. Thus, in [13] another type of fault-tolerant patterns was proposed, which is guided by the number of allowed non-missing tuples inside an n-cluster rather than by maximising their relative number. It seems that techniques searching for relaxed n-cliques maximal according a density-like criteria can be proposed for multi-mode networks as well [84]. The classic definition of *biplex* can be compared with the one of the OA-biclusters as many more similar relaxations for cliques and their possible n-adic generalisations [11].

Comparison of several existing triclustering techniques based on spectral clustering (SPECTRIC), least squares approximation (TRIBOX), OAC-prime and OAC-box operators, and formal triconcepts (TRIAS) can be found in [40, 45]. In [45], the complexity of the problem of optimal triclustering cover with respect to several quality criteria is discussed; it is shown that the problem belongs to NP-complete complexity class whereas the problem of the number of such covers belongs to #P.

Formal concepts and their lattices have been used in criminal studies to find communities of criminals operating together [72]. Many more successful applications based on FCA are known as well as related models and techniques [73, 74]. A comprehensive introduction to FCA can be found in the recent book [26] and application-oriented tutorial [32].

8   Conclusions

We have proposed a scalable technique for community detection in n-mode networks (where nodes are normally connected by hyperedges in case of $n > 2$). The approach welcomes improvements and may benefit from fine tuning and efficient filtering criteria in order to increase the scalability at the stage of density calculation and guarantee high-quality of the found communities. We consider several directions for such improvements: efficient hashing for elimination of duplicate patterns, strategies for approximate density calculation and selection of meaningful n-clusters as well as theoretical justification of choosing good thresholds for minimal density of n-clusters.

The proposed technique can also be compared with other existing approaches like fault-tolerant n-concepts [13] and with possible multimodal extensions of the existing ones like different techniques for relaxed cliques [84], variations of bicliques [68], or higher-order extensions of modularity-based criteria [66].

Since we have only showcased several relevant examples to community detection in multi-mode networks, validation of the method for analysing similar cases requires domain expert feedback, for example, by a sociologist practitioner.

Acknowledgements We would like to thank our colleagues Rakesh Agrawal, Loïc Cerf, Vincent Duquenne, Santo Fortunato, Bernhard Ganter, Jean-François Boulicaut, Mehdi Kaytoue, Boris Mirkin, Amedeo Napoli, Lhouri Nourine, Engelbert Mephu-Nguifo, Sergei Kuznetsov, Rokia Missaoui, Sergei Obiedkov, Camille Roth, Takeaki Uno, Stanley Wasserman, and Leonid Zhukov for their inspirational discussions or a piece of advice, which directly or implicitly influenced this study. We are grateful to our colleagues from the Laboratory for Internet Studies for their piece of advice as well. The study was implemented in the framework of the Basic Research Program at the National Research University Higher School of Economics in 2016 and 2017 and in the Laboratory of Intelligent Systems and Structural Analysis. The first author has also been supported by Russian Foundation for Basic Research.

Appendix: Experiments with One-Mode Networks

Table 9 Florentine family 1: 16×16, 58 edges

ρ	Covered concepts	Unique biclusters	Biclusters	Fraction of covered concepts
0	43	58	58	1.00
0.05	43	58	58	1.00
0.1	43	58	58	1.00
0.15	43	58	58	1.00
0.2	43	58	58	1.00
0.25	43	58	58	1.00
0.3	43	58	58	1.00
0.35	43	58	58	1.00
0.4	43	57	57	1.00
0.45	43	53	53	1.00
0.5	43	47	47	1.00
0.55	43	40	40	1.00
0.6	37	31	31	0.86
0.65	33	28	28	0.77
0.7	29	19	19	0.67
0.75	29	19	19	0.67
0.8	11	8	8	0.26
0.85	9	6	6	0.21
0.9	5	5	5	0.12
0.95	5	5	5	0.12
1	5	5	5	0.12

Table 10 Florentine family
2: 16×16, 46 edges

ρ	Covered concepts	Unique biclusters	Biclusters	Fraction of covered concepts
0	27	46	46	1.00
0.05	27	46	46	1.00
0.1	27	46	46	1.00
0.15	27	46	46	1.00
0.2	27	46	46	1.00
0.25	27	46	46	1.00
0.3	27	46	46	1.00
0.35	27	46	46	1.00
0.4	27	46	46	1.00
0.45	27	46	46	1.00
0.5	27	44	44	1.00
0.55	27	43	43	1.00
0.6	27	41	41	1.00
0.65	27	41	41	1.00
0.7	25	26	26	0.93
0.75	23	22	22	0.85
0.8	23	19	19	0.85
0.85	17	14	14	0.63
0.9	12	12	12	0.44
0.95	10	10	10	0.37
1	10	10	10	0.37

Table 11 Hi-tech: 36×36,
218 edges

ρ	Covered concepts	Unique biclusters	Biclusters	Fraction of covered concepts
0	191	218	218	1.00
0.05	191	218	218	1.00
0.1	191	218	218	1.00
0.15	191	218	218	1.00
0.2	191	218	218	1.00
0.25	191	218	218	1.00
0.3	191	218	218	1.00
0.35	191	213	213	1.00
0.4	191	198	198	1.00
0.45	191	174	174	1.00
0.5	189	134	134	0.99
0.55	163	99	99	0.85
0.6	126	78	78	0.66

(continued)

Table 11 (continued)

ρ	Covered concepts	Unique biclusters	Biclusters	Fraction of covered concepts
0.65	86	49	49	0.45
0.7	65	31	31	0.34
0.75	47	22	22	0.25
0.8	28	16	16	0.15
0.85	16	13	13	0.08
0.9	16	13	13	0.08
0.95	12	12	12	0.06
1	12	12	12	0.06

Table 12 Mexican people: 35×35, 268 edges

ρ	Covered concepts	Unique biclusters	Biclusters	Fraction of covered concepts
0	373	268	268	1.00
0.05	373	268	268	1.00
0.1	373	268	268	1.00
0.15	373	268	268	1.00
0.2	373	268	268	1.00
0.25	373	266	266	1.00
0.3	373	260	260	1.00
0.35	373	247	247	1.00
0.4	373	225	225	1.00
0.45	371	189	189	0.99
0.5	360	151	151	0.97
0.55	348	119	119	0.93
0.6	298	69	69	0.80
0.65	211	45	45	0.57
0.7	141	24	24	0.38
0.75	86	15	15	0.23
0.8	17	5	5	0.05
0.85	13	4	4	0.03
0.9	1	1	1	0.00
0.95	1	1	1	0.00
1	1	1	1	0.00

References

1. Agrawal, R., Gehrke, J., Gunopulos, D., Raghavan, P.: Automatic subspace clustering of high dimensional data. Data Min. Knowl. Discov. **11**(1), 5–33 (2005). doi:10.1007/s10618-005-1396-1. http://dx.doi.org/10.1007/s10618-005-1396-1
2. Akhmatnurov, M., Ignatov, D.I.: Context-aware recommender system based on boolean matrix factorisation. In: Proceedings of the Twelfth International Conference on Concept Lattices and Their Applications, pp. 99–110, Clermont-Ferrand, 13–16 October 2015. http://ceur-ws.org/Vol-1466/paper08.pdf

3. Barabási, A.: Network Science. Cambridge University Press, Cambridge (2016)
4. Barber, M.J.: Modularity and community detection in bipartite networks. Phys. Rev. E **76**, 066102 (2007). doi:10.1103/PhysRevE.76.066102. http://link.aps.org/doi/10.1103/PhysRevE. 76.066102
5. Belohlávek, R., Glodeanu, C.V., Vychodil, V.: Optimal factorization of three-way binary data using triadic concepts. Order **30**(2), 437–454 (2013). doi:10.1007/s11083-012-9254-4. http:// dx.doi.org/10.1007/s11083-012-9254-4
6. Belohlávek, R., Trnecka, M.: From-below approximations in boolean matrix factorization: Geometry and new algorithm. J. Comput. Syst. Sci. **81**(8), 1678 – 1697 (2015). doi:http://dx.doi.org/10.1016/j.jcss.2015.06.002. http://www.sciencedirect.com/ science/article/pii/S002200001500063X
7. Belohlávek, R., Vychodil, V.: Discovery of optimal factors in binary data via a novel method of matrix decomposition. J. Comput. Syst. Sci. **76**(1), 3–20 (2010). doi:10.1016/j.jcss.2009.05.002. http://dx.doi.org/10.1016/j.jcss.2009.05.002
8. Berlingerio, M., Coscia, M., Giannotti, F., Monreale, A., Pedreschi, D.: Multidimensional networks: foundations of structural analysis. World Wide Web **16**(5), 567–593 (2013). doi:10.1007/s11280-012-0190-4. http://dx.doi.org/10.1007/s11280-012-0190-4
9. Berlingerio, M., Pinelli, F., Calabrese, F.: Abacus: frequent pattern mining-based community discovery in multidimensional networks. Data Min. Knowl. Discov. **27**(3), 294–320 (2013). doi:10.1007/s10618-013-0331-0. http://dx.doi.org/10.1007/s10618-013-0331-0
10. Bohman, L.: Bringing the owners back in: an analysis of a 3-mode interlock network. Soc. Netw. **34**(2), 275 – 287 (2012). doi:http://dx.doi.org/10.1016/j.socnet.2012.01.005. //www. sciencedirect.com/science/article/pii/S037887331200007X
11. Borgatti, S.P., Everett, M.G.: Network analysis of 2-mode data. Soc. Netw. **19**(3), 243 – 269 (1997). doi:http://dx.doi.org/10.1016/S0378-8733(96)00301-2. //www.sciencedirect.com/ science/article/pii/S0378873396003012
12. Cerf, L., Besson, J., Robardet, C., Boulicaut, J.: Closed patterns meet n-ary relations. TKDD **3**(1), 3:1–3:36 (2009). doi:10.1145/1497577.1497580. http://doi.acm.org/10.1145/1497577. 1497580
13. Cerf, L., Besson, J., Nguyen, K., Boulicaut, J.: Closed and noise-tolerant patterns in n-ary relations. Data Min. Knowl. Discov. **26**(3), 574–619 (2013). doi:10.1007/s10618-012-0284-8. http://dx.doi.org/10.1007/s10618-012-0284-8
14. Chatterjee, S., Bhattacharyya, M.: Judgment analysis of crowdsourced opinions using biclustering. Inf. Sci. **375**, 138–154 (2017). doi:10.1016/j.ins.2016.09.036. http://dx.doi.org/10. 1016/j.ins.2016.09.036
15. Cichocki, A., Lee, N., Oseledets, I.V., Phan, A.H., Zhao, Q., Mandic, D.P.: Tensor networks for dimensionality reduction and large-scale optimization: Part 1 low-rank tensor decompositions. Found. Trends Mach. Learn. **9**(4–5), 249–429 (2016). doi:10.1561/2200000059. http://dx.doi. org/10.1561/2200000059
16. Codocedo, V., Napoli, A.: Lattice-based biclustering using partition pattern structures. In: ECAI 2014 - 21st European Conference on Artificial Intelligence, 18–22 August 2014, Prague - Including Prestigious Applications of Intelligent Systems (PAIS 2014), pp. 213–218 (2014). doi:10.3233/978-1-61499-419-0-213. http://dx.doi.org/10.3233/978-1-61499-419-0-213
17. Davis A., B.B.G., Gardner, M.R.: Deep South. The University of Chicago Press, Chicago (1941)
18. Doreian, P., Batagelj, V., Ferligoj, A.: Generalized blockmodeling of two-mode network data. Soc. Netw. **26**(1), 29–53 (2004). doi:http://dx.doi.org/10.1016/j.socnet.2004.01.002. //www. sciencedirect.com/science/article/pii/S0378873304000036
19. Duquenne, V.: Lattice analysis and the representation of handicap associations. Soc. Netw. **18**(3), 217–230 (1996). doi:10.1016/0378-8733(95)00274-X. http://www.sciencedirect.com/ science/article/pii/037887339500274X
20. Fararo, T.J., Doreian, P.: Tripartite structural analysis: generalizing the Breiger-Wilson formalism. Soc. Netw. **6**(2), 141–175 (1984). doi:http://dx.doi.org/10.1016/0378-8733(84)90015-7. http://www.sciencedirect.com/science/article/pii/0378873384900157

21. Fortunato, S.: Community detection in graphs. Phys. Rep. **486**(3–5), 75–174 (2010). doi:http://dx.doi.org/10.1016/j.physrep.2009.11.002. http://www.sciencedirect.com/science/article/pii/S0370157309002841

22. Freeman, L.: Finding social groups: a meta-analysis of the southern women data. In: Dynamic Social Network Modeling and Analysis: Workshop Summary and Papers, pp. 39–97. National Academy Press, Washington, DC (2003)

23. Freeman, L.C., White, D.R.: Using galois lattices to represent network data. Sociol. Methodol. **23**, 127–146 (1993)

24. Freeman, L.C.: Cliques, galois lattices, and the structure of human social groups. Soc. Netw. **18**, 173–187 (1996)

25. Ganter, B., Kuznetsov, S.O.: Pattern structures and their projections. In: Conceptual Structures: Broadening the Base, Proceedings of the 9th International Conference on Conceptual Structures, ICCS 2001, pp. 129–142, Stanford, CA, 30 July–3 August 2001. doi:10.1007/3-540-44583-8_10. http://dx.doi.org/10.1007/3-540-44583-8_10

26. Ganter, B., Obiedkov, S.A.: Conceptual Exploration. Springer, Heidelberg (2016). doi:10.1007/978-3-662-49291-8. http://dx.doi.org/10.1007/978-3-662-49291-8

27. Ganter, B., Wille, R.: Formal Concept Analysis: Mathematical Foundations, 1st edn. Springer, New York (1999)

28. Gnatyshak, D., Ignatov, D.I., Semenov, A., Poelmans, J.: Gaining insight in social networks with biclustering and triclustering. In: BIR, pp. 162–171 (2012)

29. Gnatyshak, D., Ignatov, D.I., Kuznetsov, S.O., Nourine, L.: A one-pass triclustering approach: Is there any room for big data? In: Proceedings of the Eleventh International Conference on Concept Lattices and Their Applications, pp. 231–242, Košice, 7–10 October 2014. http://ceur-ws.org/Vol-1252/cla2014_submission_26.pdf

30. Hacene, M.R., Huchard, M., Napoli, A., Valtchev, P.: Relational concept analysis: mining concept lattices from multi-relational data. Ann. Math. Artif. Intell. **67**(1), 81–108 (2013). doi:10.1007/s10472-012-9329-3. http://dx.doi.org/10.1007/s10472-012-9329-3

31. Hartigan, J.A.: Direct clustering of a data matrix. J. Am. Stat. Assoc. **67**(337), 123–129 (1972). doi:10.2307/2284710. http://dx.doi.org/10.2307/2284710

32. Ignatov, D.I.: Introduction to formal concept analysis and its applications in information retrieval and related fields. In: Information Retrieval - 8th Russian Summer School, RuSSIR 2014, pp. 42–141, Nizhniy, Novgorod, 18–22 August 2014. Revised Selected Papers (2014). doi:10.1007/978-3-319-25485-2_3. http://dx.doi.org/10.1007/978-3-319-25485-2_3

33. Ignatov, D.I.: Towards a closure operator for enumeration of maximal tricliques in tripartite hypergraphs. CoRR **abs/1602.07267** (2016). http://arxiv.org/abs/1602.07267

34. Ignatov, D.I., Kornilov, D.: RAPS: a recommender algorithm based on pattern structures. In: Proceedings of the 4th International Workshop "What Can FCA Do for Artificial Intelligence?", FCA4AI 2015, co-located with the International Joint Conference on Artificial Intelligence (IJCAI 2015), pp. 87–98, Buenos Aires, 25 July 2015. http://ceur-ws.org/Vol-1430/paper9.pdf

35. Ignatov, D.I., Kuznetsov, S.O.: Concept-based recommendations for internet advertisement. In: Belohlavek, R., Kuznetsov, S.O. (eds.) Proceedings of the CLA 2008, CEUR WS, vol. 433, pp. 157–166. Palacký University, Olomouc (2008)

36. Ignatov, D.I., Watson, B.W.: Towards a unified taxonomy of biclustering methods. In: Kuznetsov, S.O., Watson, B.W. (eds.) Proceedings of Russian and South African Workshop on Knowledge Discovery Techniques Based on Formal Concept Analysis (RuZA 2015). CEUR Workshop Proceedings, vol. 1552, pp. 23–39 (2015)

37. Ignatov, D., Kaminskaya, A., Kuznetsov, S., Magizov, R.: A concept-based biclustering algorithm. In: Proceedings of the Eight International Conference on Intelligent Information Processing (IIP-8), pp. 140–143. MAKS Press, Moscow (2010) [in Russian]

38. Ignatov, D.I., Kuznetsov, S.O., Magizov, R.A., Zhukov, L.E.: From triconcepts to triclusters. In: Rough Sets, Fuzzy Sets, Data Mining and Granular Computing - Proceedings of the 13th International Conference, RSFDGrC 2011, pp. 257–264, Moscow, 25–27 June 2011. doi:10.1007/978-3-642-21881-1_41. http://dx.doi.org/10.1007/978-3-642-21881-1_41

39. Ignatov, D.I., Kuznetsov, S.O., Poelmans, J.: Concept-based biclustering for internet advertisement. In: ICDM Workshops, pp. 123–130. IEEE Computer Society, Brussels (2012)
40. Ignatov, D.I., Kuznetsov, S.O., Poelmans, J., Zhukov, L.E.: Can triconcepts become triclusters? Int. J. Gen. Syst. **42**(6), 572–593 (2013). doi:10.1080/03081079.2013.798899. http://dx.doi.org/10.1080/03081079.2013.798899
41. Ignatov, D.I., Mikhailova, M., Zakirova, A.Y., Malioukov, A.: Recommendation of ideas and antagonists for crowdsourcing platform witology. In: Information Retrieval - 8th Russian Summer School, RuSSIR 2014, pp. 276–296, Nizhniy, Novgorod, 18–22 August 2014, Revised Selected Papers (2014). doi:10.1007/978-3-319-25485-2_9. http://dx.doi.org/10.1007/978-3-319-25485-2_9
42. Ignatov, D.I., Nenova, E., Konstantinova, N., Konstantinov, A.V.: Boolean matrix factorisation for collaborative filtering: an fca-based approach. In: Artificial Intelligence: Methodology, Systems, and Applications - Proceedings of the 16th International Conference, AIMSA 2014, pp. 47–58, Varna, 11–13 September 2014. doi:10.1007/978-3-319-10554-3_5. http://dx.doi.org/10.1007/978-3-319-10554-3_5
43. Ignatov, D.I., Kaminskaya, A.Y., Konstantinova, N., Konstantinov, A.V.: Recommender system for crowdsourcing platform witology. In: 2014 IEEE/WIC/ACM International Joint Conferences on Web Intelligence (WI) and Intelligent Agent Technologies (IAT), vol. II, pp. 327–335, Warsaw, 11–14 August 2014. doi:10.1109/WI-IAT.2014.52. http://dx.doi.org/10.1109/WI-IAT.2014.52
44. Ignatov, D.I., Kaminskaya, A.Y., Konstantinova, N., Malioukov, A., Poelmans, J.: Fca-based recommender models and data analysis for crowdsourcing platform witology. In: Graph-Based Representation and Reasoning - Proceedings of the 21st International Conference on Conceptual Structures, ICCS 2014, pp. 287–292, Iaşi, 27–30 July 2014. doi:10.1007/978-3-319-08389-6_24. http://dx.doi.org/10.1007/978-3-319-08389-6_24
45. Ignatov, D.I., Gnatyshak, D.V., Kuznetsov, S.O., Mirkin, B.G.: Triadic formal concept analysis and triclustering: searching for optimal patterns. Mach. Learn. **101**(1–3), 271–302 (2015). doi:10.1007/s10994-015-5487-y. http://dx.doi.org/10.1007/s10994-015-5487-y
46. Jäschke, R., Hotho, A., Schmitz, C., Ganter, B., Stumme, G.: TRIAS–an algorithm for mining iceberg tri-lattices. In: Proceedings of the Sixth International Conference on Data Mining, ICDM '06, pp. 907–911. IEEE Computer Society, Washington, DC (2006). doi:http://dx.doi.org/10.1109/ICDM.2006.162. http://dx.doi.org/10.1109/ICDM.2006.162
47. Jelassi, M.N., Yahia, S.B., Nguifo, E.M.: Towards more targeted recommendations in folksonomies. Soc. Netw. Anal. Min. **5**(1), 68:1–68:18 (2015). doi:10.1007/s13278-015-0307-8. http://dx.doi.org/10.1007/s13278-015-0307-8
48. Jones, I., Tang, L., Liu, H.: Community discovery in multi-mode networks. In: Paliouras, G., Papadopoulos, S., Vogiatzis, D., Kompatsiaris, Y. (eds.) User Community Discovery, pp. 55–74. Springer, Cham (2015). doi:10.1007/978-3-319-23835-7_3. http://dx.doi.org/10.1007/978-3-319-23835-7_3
49. Kaytoue, M., Kuznetsov, S.O., Napoli, A., Duplessis, S.: Mining gene expression data with pattern structures in formal concept analysis. Inf. Sci. **181**(10), 1989–2001 (2011)
50. Kaytoue, M., Kuznetsov, S.O., Macko, J., Napoli, A.: Biclustering meets triadic concept analysis. Ann. Math. Artif. Intell. **70**, 55–79 (2014). doi:10.1007/s10472-013-9379-1. http://liris.cnrs.fr/publis/?id=6292
51. Krasnov, F., Vlasova, E., Yavorskiy, R.: Connectivity analysis of computer science centers based on scientific publications data for major Russian cities. In: Proceedings of the Second International Conference on Information Technology and Quantitative Management, ITQM 2014, pp. 892–899, National Research University Higher School of Economics (HSE), Moscow, 3–5 June 2014. doi:10.1016/j.procs.2014.05.341. http://dx.doi.org/10.1016/j.procs.2014.05.341
52. Krolak-Schwerdt, S., Orlik, P., Ganter, B.: Tripat: a model for analyzing three-mode binary data. In: Bock, H.H., Lenski, W., Richter, M. (eds.) Information Systems and Data Analysis, Studies in Classification, Data Analysis, and Knowledge Organization, pp. 298–307. Springer, Berlin/Heidelberg (1994). doi:10.1007/978-3-642-46808-7_27. http://dx.doi.org/10.1007/978-3-642-46808-7_27

53. Kuznetsov, S.O.: Stability as an estimate of the degree of substantiation of hypotheses derived on the basis of operational similarity. Nauchn. Tekh. Inf., Ser.2 (Autom. Doc. Math. Ling.) **12**, 21 – 29 (1990)
54. Kuznetsov, S.O.: On stability of a formal concept. Ann. Math. Artif. Intell. **49**(1–4), 101–115 (2007)
55. Kuznetsov, S.O., Ignatov, D.: Concept stability for constructing taxonomies of web-site users,. In: Obiedkov, S., Roth, C. (eds.) Proceedings of ICFCA 2007 Satellite Workshop on Social Network Analysis and Conceptual Structures: Exploring Opportunities, pp. 19–24. Clermont-Ferrand (2007)
56. Kuznetsov, S.O., Obiedkov, S.A., Roth, C.: Reducing the representation complexity of lattice-based taxonomies. In: Conceptual Structures: Knowledge Architectures for Smart Applications, Proceedings of the 15th International Conference on Conceptual Structures, ICCS 2007, pp. 241–254, Sheffield, 22–27 July 2007. doi:10.1007/978-3-540-73681-3_18. http://dx.doi.org/10.1007/978-3-540-73681-3_18
57. Latapy, M., Magnien, C., Vecchio, N.D.: Basic notions for the analysis of large two-mode networks. Soc. Netw. **30**(1), 31 – 48 (2008). doi:10.1016/j.socnet.2007.04.006. http://www.sciencedirect.com/science/article/pii/S0378873307000494
58. Lehmann, F., Wille, R.: A triadic approach to formal concept analysis. In: Proceedings of the Third International Conference on Conceptual Structures: Applications, Implementation and Theory, pp. 32–43. Springer, London (1995). http://dl.acm.org/citation.cfm?id=645488.656867
59. Lijffijt, J., Spyropoulou, E., Kang, B., Bie, T.D.: P-n-rminer: a generic framework for mining interesting structured relational patterns. Int. J. Data Sci. Anal. **1**(1), 61–76 (2016). doi:10.1007/s41060-016-0004-3. http://dx.doi.org/10.1007/s41060-016-0004-3
60. Liu, X., Murata, T.: Evaluating community structure in bipartite networks. In: Elmagarmid, A.K., Agrawal, D. (eds.) SocialCom/PASSAT, pp. 576–581. IEEE Computer Society, Washington, DC (2010)
61. Metzler, S., Miettinen, P.: Clustering boolean tensors. Data Min. Knowl. Discov. **29**(5), 1343–1373 (2015). doi:10.1007/s10618-015-0420-3. http://dx.doi.org/10.1007/s10618-015-0420-3
62. Miettinen, P.: Boolean tensor factorizations. In: 11th IEEE International Conference on Data Mining, ICDM 2011, pp. 447–456, Vancouver, BC, 11–14 December 2011. doi:10.1109/ICDM.2011.28. http://dx.doi.org/10.1109/ICDM.2011.28
63. Mirkin, B.: Mathematical Classification and Clustering. Kluwer, Dordrecht (1996)
64. Mirkin, B.G., Kramarenko, A.V.: Approximate bicluster and tricluster boxes in the analysis of binary data. In: Proceedings of the 13th International Conference on Rough Sets, Fuzzy Sets, Data Mining and Granular Computing, RSFDGrC'11, pp. 248–256. Springer, Berlin/Heidelberg (2011). http://dl.acm.org/citation.cfm?id=2026782.2026831
65. Mohr, J.W., Duquenne, V.: The Duality of Culture and Practice: Poverty Relief in New York City, 1888–1917. Theory Soc. **26**(2/3), 305–356 (1997). Special Double Issue on New Directions in Formalization and Historical Analysis
66. Murata, T.: Detecting communities from tripartite networks. In: Rappa, M., Jones, P., Freire, J., Chakrabarti, S. (eds.) WWW, pp. 1159–1160. ACM, New York (2010)
67. Newman, M.E.J.: Scientific collaboration networks. II. shortest paths, weighted networks, and centrality. Phys. Rev. E **64**, 016,132 (2001). doi:10.1103/PhysRevE.64.016132. http://link.aps.org/doi/10.1103/PhysRevE.64.016132
68. Nussbaum, D., Pu, S., Sack, J., Uno, T., Zarrabi-Zadeh, H.: Finding maximum edge bicliques in convex bipartite graphs. Algorithmica **64**(2), 311–325 (2012). doi:10.1007/s00453-010-9486-x. http://dx.doi.org/10.1007/s00453-010-9486-x
69. Opsahl, T.: Triadic closure in two-mode networks: redefining the global and local clustering coefficients. Soc. Netw. **34** (2011). doi:10.1016/j.socnet.2011.07.001. http://www.sciencedirect.com/science/article/pii/S0378873311000360 (in press)
70. Padilha, V.A., Campello, R.J.G.B.: A systematic comparative evaluation of biclustering techniques. BMC Bioinf. **18**(1), 55:1–55:25 (2017). doi:10.1186/s12859-017-1487-1. http://dx.doi.org/10.1186/s12859-017-1487-1

71. Papalexakis, E.E., Faloutsos, C., Sidiropoulos, N.D.: Tensors for data mining and data fusion: Models, applications, and scalable algorithms. ACM Trans. Intell. Syst. Technol. **8**(2), 16:1–16:44 (2016). doi:10.1145/2915921. http://doi.acm.org/10.1145/2915921
72. Poelmans, J., Elzinga, P., Ignatov, D.I., Kuznetsov, S.O.: Semi-automated knowledge discovery: identifying and profiling human trafficking. Int. J. Gen. Syst. **41**(8), 774–804 (2012). doi:10.1080/03081079.2012.721662. http://dx.doi.org/10.1080/03081079.2012.721662
73. Poelmans, J., Ignatov, D.I., Kuznetsov, S.O., Dedene, G.: Formal concept analysis in knowledge processing: a survey on applications. Expert Syst. Appl. **40**(16), 6538–6560 (2013). doi:10.1016/j.eswa.2013.05.009. http://dx.doi.org/10.1016/j.eswa.2013.05.009
74. Poelmans, J., Kuznetsov, S.O., Ignatov, D.I., Dedene, G.: Formal concept analysis in knowledge processing: a survey on models and techniques. Expert Syst. Appl. **40**(16), 6601–6623 (2013). doi:10.1016/j.eswa.2013.05.007. http://dx.doi.org/10.1016/j.eswa.2013.05.007
75. Roth, C.: Generalized preferential attachment: towards realistic socio-semantic network models. In: ISWC 4th Intl Semantic Web Conference, Workshop on Semantic Network Analysis, Galway, CEUR-WS Series (ISSN 1613-0073), vol. 171, pp. 29–42 (2005)
76. Roth, C., Cointet, J.P.: Social and semantic coevolution in knowledge networks. Soc. Netw. **32**, 16–29 (2010)
77. Roth, C., Obiedkov, S.A., Kourie, D.G.: Towards concise representation for taxonomies of epistemic communities. In: Yahia, S.B., Nguifo, E.M., Belohlávek, R. (eds.) CLA. Lecture Notes in Computer Science, vol. 4923, pp. 240–255. Springer, Heidelberg (2006)
78. Roth, C., Obiedkov, S.A., Kourie, D.G.: On succinct representation of knowledge community taxonomies with formal concept analysis. Int. J. Found. Comput. Sci. **19**(2), 383–404 (2008). doi:10.1142/S0129054108005735. http://dx.doi.org/10.1142/S0129054108005735
79. Shin, K., Hooi, B., Faloutsos, C.: M-zoom: Fast dense-block detection in tensors with quality guarantees. In: Machine Learning and Knowledge Discovery in Databases - Proceedings of the European Conference, ECML PKDD 2016, Part I, pp. 264–280, Riva del Garda, 19–23 September 2016. doi:10.1007/978-3-319-46128-1_17. http://dx.doi.org/10.1007/978-3-319-46128-1_17
80. Spyropoulou, E., Bie, T.D., Boley, M.: Interesting pattern mining in multi-relational data. Data Min. Knowl. Discov. **28**(3), 808–849 (2014). doi:10.1007/s10618-013-0319-9. http://dx.doi.org/10.1007/s10618-013-0319-9
81. Tang, L., Liu, H., Zhang, J., Nazeri, Z.: Community evolution in dynamic multi-mode networks. In: Proceedings of the 14th ACM SIGKDD International Conference on Knowledge Discovery and Data Mining, pp. 677–685, Las Vegas, NV, 24–27 August 2008. doi:10.1145/1401890.1401972. http://doi.acm.org/10.1145/1401890.1401972
82. Tsymbal, A., Pechenizkiy, M., Cunningham, P.: Diversity in search strategies for ensemble feature selection. Inf. Fusion **6**(1), 83–98 (2005)
83. Vander Wal, T.: Folksonomy coinage and definition. URL http://vanderwal.net/folksonomy.html. http://vanderwal.net/folksonomy.html (2007). Accessed 12 Mar 2012
84. Veremyev, A., Prokopyev, O.A., Butenko, S., Pasiliao, E.L.: Exact mip-based approaches for finding maximum quasi-cliques and dense subgraphs. Comput. Optim. Appl. **64**(1), 177–214 (2016). doi:10.1007/s10589-015-9804-y. http://dx.doi.org/10.1007/s10589-015-9804-y
85. Voutsadakis, G.: Polyadic concept analysis. Order **19**(3), 295–304 (2002)
86. White, D.R.: Statistical entailments and the galois lattice. Soc. Netw. **18**(3), 201–215 (1996). doi:10.1016/0378-8733(95)00273-1. http://www.sciencedirect.com/science/article/pii/0378873395002731
87. Wille, R.: The basic theorem of triadic concept analysis. Order **12**, 149–158 (1995)
88. Wu, Z., Bu, Z., Cao, J., Zhuang, Y.: Discovering communities in multi-relational networks. In: Paliouras, G., Papadopoulos, S., Vogiatzis, D., Kompatsiaris, Y. (eds.) User Community Discovery, pp. 75–95. Springer, Cham (2015). doi:10.1007/978-3-319-23835-7_4. http://dx.doi.org/10.1007/978-3-319-23835-7_4
89. Yavorsky, R.: Research challenges of dynamic socio-semantic networks. In: Ignatov, D., Poelmans, J., Kuznetsov, S. (eds.) CEUR Workshop Proceedings, CDUD'11 - Concept Discovery in Unstructured Data, vol. 757, pp. 119–122 (2011)

90. Zachary, W.W.: An information flow model for conflict and fission in small groups. J. Anthropol. Res. **33**(4), 452–473 (1977). http://www.jstor.org/stable/3629752
91. Zakhlebin, I., Semenov, A., Tolmach, A., Nikolenko, S.I.: Detecting opinion polarisation on twitter by constructing pseudo-bimodal networks of mentions and retweets. In: Information Retrieval - 9th Russian Summer School, RuSSIR 2015, pp. 169–178, Saint Petersburg, 24–28 August 2015, Revised Selected Papers (2015). doi:10.1007/978-3-319-41718-9_10. http://dx. doi.org/10.1007/978-3-319-41718-9_10
92. Zhao, L., Zaki, M.J.: Tricluster: An effective algorithm for mining coherent clusters in 3d microarray data. In: Proceedings of the ACM SIGMOD International Conference on Management of Data, pp. 694–705, Baltimore, Maryland, 14–16 June 2005. doi:10.1145/1066157.1066236. http://doi.acm.org/10.1145/1066157.1066236
93. Zhuk, R., Ignatov, D.I., Konstantinova, N.: Concept learning from triadic data. In: Proceedings of the Second International Conference on Information Technology and Quantitative Management, ITQM 2014, pp. 928–938, National Research University Higher School of Economics (HSE), Moscow, 3–5 June 2014. doi:10.1016/j.procs.2014.05.345. http://dx.doi. org/10.1016/j.procs.2014.05.345
94. Zudin, S., Gnatyshak, D.V., Ignatov, D.I.: Putting oac-triclustering on mapreduce. In: Proceedings of the Twelfth International Conference on Concept Lattices and Their Applications, pp. 47–58, Clermont-Ferrand, 13–16 October 2015. http://ceur-ws.org/Vol-1466/paper04.pdf

Acquisition of Terminological Knowledge from Social Networks in Description Logic

Francesco Kriegel

1 Introduction and Problem Description

In the last years, a rapidly increasing amount of data was collected and recorded in so-called *triple stores*. Basically, those triple stores are databases of a special kind, allowing for storing data in the form of triples (s, p, o) which express that the subject s is related to the object o via the (binary) predicate p. For example, it is possible to say that an individual x is a human by means of the triple $(x, \mathtt{rdf:type}, \mathtt{some\text{-}namespace:human})$. As another example, with the triple $(x, \mathtt{foaf:hasFriend}, y)$ we can denote that individual x is a friend of the individual y. The vocabulary used in the triples can be freely chosen such that it best fits the application's needs. Please note that there are plenty of vocabularies available, which could be used without requiring to invent one's own vocabulary from scratch. The most famous examples are, of course, the vocabularies from *RDF/RDFS* and *OWL* which allow for the expression of very basic and logical facts. Further vocabularies specifically tailored to certain use cases are, e.g., *Friend-of-a-Friend (FOAF)* and others. It is easy to see that those triple datasets can also be represented as *labeled directed graphs*, the vertices of which are the elements occurring as subjects or objects, and each triple (s, p, o) induces an edge from s to o with label p. Labels of vertices are induced by triples of the form $(s, \mathtt{rdf:type}, c)$, and in particular for each such triple, the vertex s is labeled with c.

The *Web Ontology Language (OWL)* was founded in 2004 as an improvement of the *Resource Description Framework (RDF)* and the corresponding *RDF Schema (RDFS)*. OWL and its successor OWL2 have various dialects providing different expressibility and complexity such that always one can be chosen that best fits the user's purpose. Most of the dialects, and in particular the dialects OWL DL, OWL2

F. Kriegel (✉)
Institute of Theoretical Computer Science, Technische Universität Dresden, Dresden, Germany
e-mail: francesco.kriegel@tu-dresden.de

© Springer International Publishing AG 2017
R. Missaoui et al. (eds.), *Formal Concept Analysis of Social Networks*,
Lecture Notes in Social Networks, DOI 10.1007/978-3-319-64167-6_5

DL, and OWL2 EL, have a strong logical underpinning by means of *Description Logics (DLs)*. DLs are a family of logical languages for knowledge representation and reasoning, for which the decidability and complexity of common reasoning problems are widely explored. Those reasoning tasks allow for the deduction of implicit knowledge from explicitly given facts and axioms, and a vast amount of algorithms for solving those reasoning problems were developed, optimized, and implemented—the most popular ones are the tableaux algorithms and the completion algorithms.

An interesting problem in the field of Description Logics is the problem of learning, a specific instance of which is the acquisition of terminological knowledge from a given set of assertional facts. So far there are several techniques for achieving this, and some of them utilize the algorithmic solutions of the problem of computing implication bases in the field of *Formal Concept Analysis*, or utilize the *Attribute Exploration* algorithm that is capable of handling incomplete data by incorporating an expert in the domain of interest which is able to answer questions correctly and thus enables the algorithm to process axioms the validity of which is either not answerable within the input dataset, or is not refuted due to the non-existence of a counterexample. A famous work in this direction was published by Baader and Distel [2, 3, 16] who generalized the computation, or exploration, respectively, of implication bases for formal contexts to the computation, or exploration, respectively, of bases of concept inclusions (CIs) valid in a given interpretation and expressible in the description logic $\mathcal{EL}^\perp$. Furthermore, Borchmann [10, 11] defined the notion of *confidence* of a CI within an interpretation, a measure indicating which fraction of the individuals in the interpretation fulfill a certain CI. He then developed a technique for the construction of a base of CIs the confidence of which exceeds a pre-defined threshold in [0, 1]. His work is particularly useful for datasets occuring in practical use cases where it cannot be ruled out that there is some noise, i.e., errors, in the dataset to be analyzed. Borchmann then also investigated and constituted an explorative method for the axiomatization of confident CIs, which also needs an interpretation as input, and furthermore an expert that is capable of correctly answering questions in the domain of interest.

We consider social networks that are encoded as *description graphs*, i.e., as directed graphs the vertices and edges of which are labeled. The aim is to extract terminological axioms, so-called *concept inclusions*, from the graph in order to describe the logical structure of the social network. Furthermore, we assume that the underlying graph to be analyzed is complete and error-free, i.e., fully describes all persons and entities in the social network as well as their connections. It is straightforward that description graphs and interpretations are isomorphic—we will later elaborate on this fact. In particular, we consider a social network that is given in form of an interpretation $\mathcal{I}$, which we indeed may assume for the aforementioned reason. Our aim now is to formulate terminological axioms that are valid in $\mathcal{I}$, i.e., we are searching for CIs $C \sqsubseteq D$ that are valid in $\mathcal{I}$. Furthermore, we shall do this in a complete manner. However, it is easy to see that the number of concept inclusions that are expressible over a given signature is infinite; and in case of a restricted role depth and a finite signature there are only finitely many concept inclusions. By some simple observations, one can verify that the number of concept descriptions with a role depth of $\delta + 1$ is exponential in the number of concept descriptions with a role depth of δ. Consequently, it would certainly not be a good idea to enumerate *all* valid concept inclusions of $\mathscr{I}$. We should rather try to find a base for the valid CIs of $\mathscr{I}$, as it has been first investigated by Baader and Distel in [2, 16] with respect to *greatest fixpoint semantics*, and later by Borchmann, Distel, and Kriegel, in [12] with respect to *descriptive semantics* (the *default* semantics). A *base of CIs* for $\mathscr{I}$ is a TBox $\mathscr{B}$ such that for each concept inclusion $C \sqsubseteq D$, $\mathscr{I} \models C \sqsubseteq D$ if, and only if, $\mathscr{B} \models C \sqsubseteq D$. A slight generalization of the notion of a base for an interpretation has been introduced in [29], which allows for the incorporation of existing knowledge.

In this chapter we in particular provide a generalization of the aforementioned means for constructing bases of CIs in the more expressive description logic $\mathscr{M}\mathscr{H}$, and also demonstrate how the technique can be applied to social graphs. This chapter is structured as follows. In Sect. 2 the notion of a *social graph* is defined, and it is shown that the data model of *Facebook* induces a social graph. Section 3 gives a short introduction to the *Web Ontology Language (OWL)*, and the following Sect. 4 presents the description logic $\mathscr{M}\mathscr{H}$ which is a monotonous fragment of the DL $\mathscr{S}\mathscr{R}\mathscr{O}\mathscr{I}\mathscr{Q}$ underlying the second version of OWL. Then in Sect. 5 we investigate the lattice induced by the $\mathscr{M}$-concept descriptions. Section 6 gives a brief introduction to *Formal Concept Analysis*. In Sect. 7 we show that each interpretation in the description logic $\mathscr{M}\mathscr{H}$ induces a Galois connection between the set of $\mathscr{M}\mathscr{H}$-concept descriptions and the powerset of the interpretation's domain; in particular Sect. 8 justifies the existence of the aforementioned Galois connection by providing a construction for so-called *role-depth-bounded model-based most specific concept descriptions* in the DL $\mathscr{M}$. Section 9 generalizes the notion of a *concept lattice* from formal contexts to $\mathscr{M}\mathscr{H}$-interpretations. Furthermore, Sect. 10 presents an important connection between *Formal Concept Analysis* and $\mathscr{M}\mathscr{H}$-interpretations, which is then utilized in Sect. 11 to develop a construction method for knowledge bases of $\mathscr{M}\mathscr{H}$-interpretations. Eventually, Sect. 12 gives a short overview on description logics the expressivity of which is below $\mathscr{M}\mathscr{H}$ and that may also be used as a language for axiomatizing terminological knowledge. The chapter closes with Sect. 13.

2 Social Networks and Social Graphs

A *social graph* is a directed graph the vertices and edges of which are labeled. The vertices represent the entities, e.g., persons, events, messages, etc., and the edges represent relationships between the entities, e.g., friendship between persons, attendance of a person to an event, a person liking a message, etc. Formally, we describe social networks as follows. First, fix a set N_V of vertex labels as well as a set N_E of edge labels. Then, a *social graph* over (N_V, N_E) is a tuple $\mathscr{G} := (V, E, L_V, L_E)$ where

Fig. 1 An exemplary social
graph

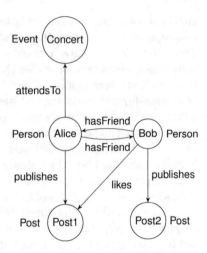

1. V is a set of vertices,
2. $E \subseteq V \times V$ is a set of directed edges,
3. $L_V: V \to \wp(N_V)$ is a vertex labeling function, and
4. $L_E: E \to \wp(N_E)$ is an edge labeling function.

A toy example of a social graph is shown in Fig. 1. It contains two persons, Alice and Bob, which are friends. Furthermore, Alice attends a concert and publishes a message which Bob likes. Bob publishes a message, too.

As an exemplary social network we consider *Facebook* [19], which is the most popular social network as of 2017. It has been founded by *Mark Zuckerberg*, and its website was launched in 2004. In the beginning it was limited to students from Harvard, but was later opened stepwise to a broader audience. In 2006 everybody with an age of at least 13 was allowed to create an account on Facebook. Since its beginning it has successfully evolved to a networking platform, which allows its users to publish messages, share photos, etc., and interact with each other, e.g., by liking other's activities, communicating with private messages, connecting by (digital) friendship, etc. Facebook's data is available via the *Facebook Graph API*, cf. [20]. Its data model fits well for our use case—it is accessible as a directed graph with labeled vertices and edges. In general the Facebook graph consists of nodes, edges, and fields. The nodes represent entities, like persons, photos, comments, events, etc.; the edges represent connections between the entities, e.g., an edge could link a photo to a person, or express that two persons are virtual friends; the fields represent information about the entities, e.g., a person's name, a person's birthday, the publish date of a comment, etc. In terms of description logics, those field values can be expressed by appropriate values in concrete domains. We will not go into detail here, and rather refer the interested reader to [20].

3   The Web Ontology Language (OWL)

The *Web Ontology Language (OWL)* was introduced in its first version in 2004 as an extension of the *Resource Description Framework (RDF)* and *RDF Schema (RDFS)* in order to provide a well-founded semantics and to increase the expressibility of the language. There were some language constructs expressible in RDF/RDFS leading to inconsistencies or undecidability that are not expressible in OWL anymore, i.e., OWL resolved this issue. Later in 2009, a more expressive second version *OWL2* was founded.

However, RDF was not fully replaced, but remained a storage format for OWL, besides other formats, e.g., XML, Manchester Syntax, etc. A new vocabulary was defined, which allowed for the expression of the language constructs of OWL, e.g., the predicate `owl:isA` for assigning types to individuals (similar to `rdf:type`), the predicate `owl:subClassOf` for expressing subclass relationships, etc. For a full reference, the reader is referred to [46]—in the sequel of this chapter we only consider some of the provided language constructs. In particular, we will leave out concrete domains, disjunctions and negations, and others. Additionally, plenty of information including interesting examples and use cases can be found in the book [26] of Hitzler, Krötzsch, and Rudolph. OWL and its dialects are used for the *Semantic Web* and for *Linked Data*, e.g., in the medical domain (*SNOMED* ontology), and in *DBpedia* as well as *Wikidata* (structured machine-readable derivations of *Wikipedia*).

The logical underpinning of OWL and some of its dialects is provided by *Description Logics (DLs)*, which are a family of conceptual languages suitable for knowledge representation and reasoning that have a strong logical foundation for which the decidability and complexity of common reasoning problems is widely explored. In particular, the reasoning tasks allow for deduction of implicit knowledge from explicitly stated facts and axioms, and plenty of appropriate algorithms were developed and implemented, e.g., tableaux algorithms and completion algorithms. In particular, the full first version of the Web Ontology Language corresponds to the description logic $\mathcal{SHOIN}$, and the full second version of the Web Ontology Language is covered by the description logic $\mathcal{SROIQ}$. In the next Sect. 4, we shall focus on (a fragment of) the description logic $\mathcal{SROIQ}$, which is suitable for terminological learning, i.e., which allows for a certain degree of abstraction and not only rewrites given assertional data into terminological axioms. In particular, this implies that we shall not make use of neither negation, nor disjunction, nor nominals, nor other constructors that can emulate the aforementioned.

4   The Description Logic $\mathcal{MH}$

This section presents the description logic $\mathcal{ALQ}^{\geq}\mathcal{N}^{\leq}(\mathsf{Self})$, which is a fragment of $\mathcal{SROIQ}$, and allows for conjunctions, primitive negations, value restrictions, qualified at-least restrictions, unqualified at-most restrictions, and existential self restrictions. Furthermore, we will not focus on the implementation details of OWL, and do not present any of the different syntaxes of OWL, but rather use the theoretical notations that are used in the field of description logics. The considered description logic $\mathcal{ALQ}^{\geq}\mathcal{N}^{\leq}(\text{Self})$ is abbreviated as $\mathcal{M}$, which encodes the *monotonicity* of all allowed constructors.

Consider a finite *signature* $\Sigma := (N_C, N_R)$, that is, N_C is a finite set of *concept names*, and N_R is a finite set of *role names*. Then an $\mathcal{M}$-*concept description* over Σ can be constructed according to the following inductive rule where $A \in N_C$, $r \in N_R$, and $n \in \mathbb{N}$.

$$C ::= \bot \mid \top \mid A \mid \neg A \mid C \sqcap C \mid \forall r.C \mid \exists \geq n.r.C \mid \exists \leq n.r \mid \exists r.\,\text{Self}$$

The semantics are *model-theoretic*, that is, they are defined by means of so-called *interpretations*. An *interpretation* $\mathcal{I}$ over $\Sigma = (N_C, N_R)$ is a pair $(\Delta^{\mathcal{I}}, \cdot^{\mathcal{I}})$ consisting of a non-empty set $\Delta^{\mathcal{I}}$ which is called *domain*, and an *extension function* $\cdot^{\mathcal{I}} : N_C \cup N_R \rightarrow \wp(\Delta^{\mathcal{I}}) \cup \wp(\Delta^{\mathcal{I}} \times \Delta^{\mathcal{I}})$ that maps concept names $A \in N_C$ to subsets $A^{\mathcal{I}} \subseteq \Delta^{\mathcal{I}}$, and role names $r \in N_R$ to binary relations $r^{\mathcal{I}} \subseteq \Delta^{\mathcal{I}} \times \Delta^{\mathcal{I}}$. The extension function is then canonically extended to all $\mathcal{M}$-concept descriptions according to the following recursive definitions.

$$\bot^{\mathcal{I}} := \emptyset$$

$$\top^{\mathcal{I}} := \Delta^{\mathcal{I}}$$

$$(\neg A)^{\mathcal{I}} := \Delta^{\mathcal{I}} \setminus A^{\mathcal{I}}$$

$$(C \sqcap D)^{\mathcal{I}} := C^{\mathcal{I}} \cap D^{\mathcal{I}}$$

$$(\forall r.C)^{\mathcal{I}} := \{ d \in \Delta^{\mathcal{I}} \mid \forall e \in \Delta^{\mathcal{I}} : (d,e) \in r^{\mathcal{I}} \text{ implies } e \in C^{\mathcal{I}} \}$$

$$(\exists \geq n.r.C)^{\mathcal{I}} := \{ d \in \Delta^{\mathcal{I}} \mid |\{ e \in \Delta^{\mathcal{I}} \mid (d,e) \in r^{\mathcal{I}} \text{ and } e \in C^{\mathcal{I}} \}| \geq n \}$$

$$(\exists \leq n.r)^{\mathcal{I}} := \{ d \in \Delta^{\mathcal{I}} \mid |\{ e \in \Delta^{\mathcal{I}} \mid (d,e) \in r^{\mathcal{I}} \}| \leq n \}$$

$$(\exists r.\,\text{Self})^{\mathcal{I}} := \{ d \in \Delta^{\mathcal{I}} \mid (d,d) \in r^{\mathcal{I}} \}$$

Of course, we may emulate *existential restrictions*, the expressibility of which is symbolized by the letter $\mathcal{E}$ within the description logic's name, by using the abbreviation $\exists r.C := \exists \geq 1.r.C$, i.e., both $\mathcal{M}$ and $\mathcal{ME} := \mathcal{ALEQ}^{\geq}\mathcal{N}^{\leq}(\text{Self})$ denote essentially the same logic. It is readily verified that the following equation for the extension of existential restrictions is satisfied.

$$(\exists r.C)^{\mathcal{I}} = \{ d \in \Delta^{\mathcal{I}} \mid \exists e \in \Delta^{\mathcal{I}} : (d,e) \in r^{\mathcal{I}} \text{ and } e \in C^{\mathcal{I}} \}$$

Informally, the *role depth* of a concept description is defined as the maximal number of nestings of role quantifiers. More specifically, we define the *role depth* $\text{rd}(C)$ of an $\mathcal{M}$-concept description C recursively as follows.

$$\mathsf{rd}(\bot) := 0$$

$$\mathsf{rd}(\top) := 0$$

$$\mathsf{rd}(A) := 0 \qquad\qquad \text{for each concept name } A \in N_C$$

$$\mathsf{rd}(\neg A) := 0 \qquad\qquad \text{for each concept name } A \in N_C$$

$$\mathsf{rd}(C \sqcap D) := \max(\mathsf{rd}(C), \mathsf{rd}(D))$$

$$\mathsf{rd}(\forall\, r.\, C) := 1 + \mathsf{rd}(C)$$

$$\mathsf{rd}(\exists \geq n.\, r.\, C) := 1 + \mathsf{rd}(C)$$

$$\mathsf{rd}(\exists \leq n.\, r\,) := 1$$

$$\mathsf{rd}(\exists\, r.\, \mathsf{Self}) := 1$$

The set of all $\mathcal{M}$-concept descriptions over a signature Σ is symbolized as $\mathcal{M}(\Sigma)$, and for a role-depth bound $\delta \in \mathbb{N}$, we denote by $\mathcal{M}(\Sigma)\!\restriction_\delta$ the set of all $\mathcal{M}$-concept descriptions over Σ with a role depth not exceeding δ.

A *concept inclusion* (abbr. *CI*) is an expression $C \sqsubseteq D$ where both C and D are concept descriptions. A *terminological box* (abbr. *TBox*) is a finite set of concept inclusions. A CI $C \sqsubseteq D$ is valid in $\mathcal{I}$ if $C^{\mathcal{I}} \subseteq D^{\mathcal{I}}$. We then also refer to $\mathcal{I}$ as a *model* of $C \sqsubseteq D$, and denote this by $\mathcal{I} \models C \sqsubseteq D$. Furthermore, $\mathcal{I}$ is a *model* of a TBox $\mathcal{T}$, symbolized as $\mathcal{I} \models \mathcal{T}$, if each CI in $\mathcal{T}$ is valid in $\mathcal{I}$. The entailment relation is lifted to TBoxes as follows: A CI $C \sqsubseteq D$ is *entailed* by a TBox $\mathcal{T}$, denoted as $\mathcal{T} \models C \sqsubseteq D$, if each model of $\mathcal{T}$ is a model of $C \sqsubseteq D$, too. We then also say that C is *subsumed* by D with respect to $\mathcal{T}$. A TBox $\mathcal{T}$ *entails* a TBox $\mathcal{U}$, symbolized as $\mathcal{T} \models \mathcal{U}$, if $\mathcal{T}$ entails each CI in $\mathcal{U}$, or equivalently if each model of $\mathcal{T}$ is also a model of $\mathcal{U}$. Two $\mathcal{M}$-concept descriptions C and D are *equivalent* with respect to $\mathcal{T}$, and we shall write $\mathcal{T} \models C \equiv D$, if $\mathcal{T} \models \{C \sqsubseteq D, D \sqsubseteq C\}$. In case $\mathcal{T} = \emptyset$ we may omit the prefix "$\emptyset \models$". However, then we have to carefully interpret an expression $C \sqsubseteq D$—it either just denotes a concept inclusion, i.e., an axiom, without stating where it is valid; or it expresses that C is subsumed by D (w.r.t. $\emptyset$), i.e., $C^{\mathcal{I}} \subseteq D^{\mathcal{I}}$ is satisfied in all interpretations $\mathcal{I}$. An analogous hint applies to concept equivalences $C \equiv D$.

To justify the choice of the abbreviation $\mathcal{M}$ for $\mathcal{ALQ}^{\geq}\mathcal{N}^{\leq}(\mathsf{Self})$, we remark that each of the constructors is *monotonous*, i.e., it holds true that for all $\mathcal{M}$-concept descriptions C, D, E, all role names $r \in N_R$, and all natural numbers $n \in \mathbb{N}$,

$$\{C \sqsubseteq D\} \models \{C \sqcap E \sqsubseteq D \sqcap E,\ \forall r.\, C \sqsubseteq \forall r.\, D,\ \exists \geq n.\, r.\, C \sqsubseteq \exists \geq n.\, r.\, D\}.$$

A *role inclusion* (abbr. *RI*) is an expression $r \sqsubseteq s$ where $r, s \in N_R$ are role names. A *relational box* (abbr. *RBox*) is a finite set of role inclusions. For an interpretation $\mathcal{I}$, we say that $r \sqsubseteq s$ is *valid* in $\mathcal{I}$, denoted as $\mathcal{I} \models r \sqsubseteq s$, if $r^{\mathcal{I}} \subseteq s^{\mathcal{I}}$. Furthermore, an RBox $\mathcal{R}$ is *valid* in $\mathcal{I}$, symbolized as $\mathcal{I} \models \mathcal{R}$, if each role inclusion in $\mathcal{R}$ is valid in $\mathcal{I}$. In case a description logic allows for the usage of these role inclusions, then its name contains the letter $\mathcal{H}$. In what follows we are going to merely consider the description logic $\mathcal{M}\mathcal{H}$.

In order to decide entailment, the well-known *tableaux algorithm* [5, Sect. 3.4] can be utilized. It takes as input a knowledge base $(\mathcal{T}, \mathcal{A})$ consisting of a TBox and an ABox, and tries to construct a model of the knowledge base. It was shown that the tableaux algorithm is sound (i.e., the output is indeed a model), complete (i.e., if a model exists, then a model is constructed and returned), and terminates (i.e., for finite input yields a result after a finite amount of time). These are the following common reasoning problems, cf. [5, Sect. 3.2.2].

1. *Knowledge Base Consistency:* Given a knowledge base $\mathcal{K}$, is there a model of $\mathcal{K}$?
2. *Concept Satisfiability:* Given a concept description C, and a knowledge base $\mathcal{K}$, is there a model of $\mathcal{K}$ in which C has a non-empty extension?
3. *Concept Subsumption:* Given two concept descriptions C and D, and a knowledge base $\mathcal{K}$, does $\mathcal{I} \models C \sqsubseteq D$ hold true for all models $\mathcal{I}$ of $\mathcal{K}$?
4. *Concept Equivalence:* Given two concept descriptions C and D, and a knowledge base $\mathcal{K}$, does $\mathcal{I} \models C \equiv D$ hold true for all models $\mathcal{I}$ of $\mathcal{K}$?
5. *Instance Checking:* Given an individual a, a concept description C, and a knowledge base $\mathcal{K}$, does $\mathcal{K}$ entail $a \in C$?
6. *Role Instance Checking:* Given two individuals a and b, a role name r, and a knowledge base $\mathcal{K}$, does $\mathcal{K}$ entail $(a, b) \in r$?

There is a strong correspondence between interpretations and directed labeled graphs, and in particular it is easy to translate between both formalisms. We start with defining a *description graph*, which is very similar to a social graph as introduced in Sect. 2. A *description graph* over a signature (N_C, N_R) is a tuple $\mathcal{G} := (V, E, L_V, L_E)$ that satisfies the following conditions.

1. (V, E) is a directed graph, i.e., V is a set of *vertices*, and $E \subseteq V \times V$ is a set of *directed edges*,
2. $L_V: V \to \wp(N_C)$ is a *vertex labelling*, and
3. $L_E: E \to \wp(N_R)$ is an *edge labelling*.

Please note that in some works description graphs are defined to have a distinguished root vertex—however, this is not necessary for our purposes.

Each interpretation induces a directed labeled graph as follows: let $\mathcal{I} := (\Delta^{\mathcal{I}}, \cdot^{\mathcal{I}})$ be an interpretation over the signature (N_C, N_R). Then, define the description graph $\mathcal{G}(\mathcal{I}) := (V, E, L_V, L_E)$ over (N_C, N_R) that consists of the directed graph (V, E) with the components

$$V := \Delta^{\mathcal{I}},$$

$$\text{and} \quad E := \bigcup \{ r^{\mathcal{I}} \mid r \in N_R \},$$

and the corresponding labeling functions

$$L_V: \quad V \to \wp(N_C)$$
$$x \mapsto \{A \in N_C \mid x \in A^{\mathscr{I}}\},$$
$$\text{and} \quad L_E: \quad E \to \wp(N_R)$$
$$(x, y) \mapsto \{r \in N_R \mid (x, y) \in r^{\mathscr{I}}\}.$$

Note that $\mathscr{G}(\mathscr{I})$ just formalizes the natural graphical representation of interpretations as they are usually drawn in toy examples.

Vice versa, if $\mathscr{G} := (V, E, L_V, L_E)$ is a description graph over (N_C, N_R), then its induced interpretation is $\mathscr{I}(\mathscr{G}) := (\Delta^{\mathscr{I}(\mathscr{G})}, \cdot^{\mathscr{I}(\mathscr{G})})$ the components of which are defined in the following way.

$$\Delta^{\mathscr{I}(\mathscr{G})} := V,$$

$$\text{and} \quad \cdot^{\mathscr{I}(\mathscr{G})} \colon \begin{cases} A \mapsto \{x \in V \mid A \in L_V(x)\} \\ r \mapsto \{(x, y) \in E \mid r \in L_E(x, y)\}. \end{cases}$$

It is readily verified that the two transformations are mutually inverse, and this justifies that we do not have to distinguish between interpretations and description graphs (or social graphs) in the sequel of this document.

5   The Lattice of $\mathscr{M}$-Concept Descriptions

It is readily verified that the *subsumption* $\sqsubseteq$ with respect to the empty TBox $\emptyset$ constitutes a *quasi-order* on the set $\mathscr{M}(\Sigma)$ of all $\mathscr{M}$-concept descriptions over the signature $\Sigma = (N_C, N_R)$, i.e., the following conditions are satisfied.

1. $\sqsubseteq$ w.r.t. $\emptyset$ is *reflexive*, i.e., for all $\mathscr{M}$-concept descriptions C, $\emptyset \models C \sqsubseteq C$, and
2. $\sqsubseteq$ w.r.t. $\emptyset$ is *transitive*, i.e., for all $\mathscr{M}$-concept descriptions C, D, E, it holds true that $\emptyset \models C \sqsubseteq D$ and $\emptyset \models D \sqsubseteq E$ implies $\emptyset \models C \sqsubseteq E$.

Of course, then the *equivalence* $\equiv$ with respect to $\emptyset$ is an *equivalence relation*, i.e., the following statements hold true.

1. $\equiv$ w.r.t. $\emptyset$ is *reflexive*, i.e., for all $\mathscr{M}$-concept descriptions C, $\emptyset \models C \equiv C$,
2. $\equiv$ w.r.t. $\emptyset$ is *transitive*, i.e., for all $\mathscr{M}$-concept descriptions C, D, E, we have that $\emptyset \models C \equiv D$ and $\emptyset \models D \equiv E$ implies $\emptyset \models C \equiv E$, and
3. $\equiv$ w.r.t. $\emptyset$ is *symmetric*, i.e., for all $\mathscr{M}$-concept descriptions C, D, it holds true that $\emptyset \models C \equiv D$ implies $\emptyset \models D \equiv C$.

By definition it follows that it is the induced equivalence relation of $\sqsubseteq$, i.e., $\emptyset \models C \equiv D$ if, and only if, $\emptyset \models C \sqsubseteq D$ as well as $\emptyset \models D \sqsubseteq C$. Hence, the quotient of

$(\mathscr{M}(\Sigma), \sqsubseteq)$ with respect to the induced *equivalence* $\equiv$ w.r.t. $\emptyset$ is a *partially ordered set* (a *poset*). It consists of all *equivalence classes* $[C]_\equiv$ for $\mathscr{M}$-concept descriptions C, which are defined by

$$[C]_\equiv := \{ D \mid \emptyset \models C \equiv D \}.$$

Furthermore, for an equivalence class $[C]_\equiv$, we say that C is a *representative* of it. We can then define a *partial order* on the classes which is induced by the subsumption between their representatives, i.e., for all $\mathscr{M}$-concept descriptions C, D,

$$\emptyset \models [C]_\equiv \sqsubseteq [D]_\equiv \text{ if, and only if, } \emptyset \models C \sqsubseteq D.$$

This partial order enjoys all properties of a quasi-order as stated above, and furthermore is *anti-symmetric*, i.e., for all $\mathscr{M}$-concept descriptions C, D,

$$\emptyset \models [C]_\equiv \sqsubseteq [D]_\equiv \text{ and } \emptyset \models [D]_\equiv \sqsubseteq [C]_\equiv \text{ implies } [C]_\equiv = [D]_\equiv.$$

For the sake of simplicity, we will not distinguish between the equivalence classes and their representatives in the sequel of this chapter. The poset $(\mathscr{M}(\Sigma), \sqsubseteq)/{\equiv}$ is even a bounded lattice. Of course, $\bot$ is the smallest element, and $\top$ is the greatest element. It is easy to see that the (finitary) conjunction $\sqcap$ corresponds to the finitary *infimum* operation, since for all finite sets $\mathscr{C}$ of $\mathscr{M}$-concept descriptions over Σ, it holds that the conjunction $\sqcap \mathscr{C}$ is the greatest lower bound (w.r.t. $\sqsubseteq$) of all concept descriptions in $\mathscr{C}$, i.e., $\emptyset \models \sqcap \mathscr{C} \sqsubseteq C$ for all $C \in \mathscr{C}$, and for all $\mathscr{M}$-concept descriptions D with $\emptyset \models D \sqsubseteq C$ for all $C \in \mathscr{C}$, it holds true that $\emptyset \models D \sqsubseteq \sqcap \mathscr{C}$. However, what is missing is a *supremum* operation. Of course, in description logics allowing for *disjunction*, we can easily prove that the disjunction is the supremum operation. For the general case, the notion of a *smallest upper bound* is rather called *least common subsumer* in the field of description logics, and is defined as follows.

Definition 5.1 Let C, D be $\mathscr{M}$-concept descriptions over the signature Σ. Then a concept description $E \in \mathscr{M}(\Sigma)$ is called a *least common subsumer* (abbr. *LCS*) of C and D if the following conditions are fulfilled.

1. E subsumes both C and D, i.e., $\emptyset \models C \sqsubseteq E$ and $\emptyset \models D \sqsubseteq E$.
2. Whenever F is a common subsumer of C and D, then F subsumes E, i.e., for all concept descriptions $F \in \mathscr{M}(\Sigma)$, $\emptyset \models \{C \sqsubseteq F, D \sqsubseteq F\}$ implies $\emptyset \models E \sqsubseteq F$.

It follows that least common subsumers are always unique up to equivalence. Hence, we can speak of *the* LCS of two concept descriptions, and furthermore we denote it by $C \vee D$. The definition can be canonically extended to an arbitrary number of concept descriptions, and we then write $\bigvee \mathscr{C}$ for the least common subsumer of a set $\mathscr{C}$ of $\mathscr{M}$-concept descriptions over Σ. It is readily verified that the conjunction is a categorical product, cf. Fig. 2, and dually the least common subsumer is a categorical coproduct, cf. Fig. 3.

Fig. 2 The conjunction is a product in the category the objects of which are concept descriptions and the morphisms of which are subsumptions, cf. [38, p. 69]

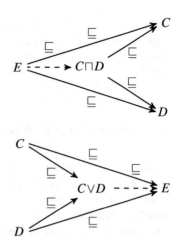

Fig. 3 The least common subsumer is a coproduct in the category the objects of which are concept descriptions and the morphisms of which are subsumptions, cf. [38, p. 63]

It was shown that least common subsumers always exist in several description logics, e.g., in $\mathscr{EL}$, $\mathscr{FLE}$, and $\mathscr{ALE}$, as shown in [4] by Baader, Küsters, and Molitor; in $\mathscr{ALQ}$ and $\mathscr{ALENR}$ as shown in [40, 41] by Mantay; in $\mathscr{ALEN}$ as shown in [33] by Küsters and Molitor; in $\mathscr{ALEHIN}_{R+}$ as shown in [18] by Donini, Colucci, Di Noia, and Di Sciascio; in $\mathscr{EL}_{\text{gfp}}$, i.e., $\mathscr{EL}$ interpreted with *greatest fixpoint semantics*, as shown in [1] by Baader; in $\mathscr{FLE}_{\text{gfp}}$ as shown in [14] by Distel; and in $\mathscr{EL}^{\perp}_{\text{gfp}}$ as shown by Distel in [16].

As a practical means for ensuring the existence of least common subsumers, we could also apply a bound on the role depth of the concept descriptions under consideration. For the case of $\mathscr{EL}^{\perp}$ this has been done in [12] by Borchmann, Distel, and Kriegel. However, this result also applies to all other description logics equipped with a bound on the role depths—in particular, we know that then for all concept descriptions C and D, there are only finitely many concept descriptions that satisfy the role depth bound, use only concept names and role names occuring in C or D, and that only include numbers in at-least or at-most restrictions not exceeding those occuring in C or D. Denote the conjunction of these three properties by $*$. Then, we can infer that

$$\emptyset \models C \vee D \equiv \bigsqcap \{E \mid E \text{ satisfies } * \text{ and } \emptyset \models \{C \sqsubseteq E, D \sqsubseteq E\}\},$$

holds true and is a well-defined formula as the set $\{E \mid E \text{ satisfies } * \text{ and } \emptyset \models \{C \sqsubseteq E, D \sqsubseteq E\}\}$ must be finite, and thus its conjunction indeed exists. Note that this is a rather theoretical argument showing the existence, but not allowing for a practical computation of least common subsumers.

It is easy to see that the equivalence $\equiv$ is compatible with both $\sqcap$ and $\vee$. In the sequel of this chapter, we shall denote this bounded lattice by $\underline{\mathscr{M}}(\Sigma) := (\mathscr{M}(\Sigma), \sqsubseteq)/{\equiv}$, and accordingly $\underline{\mathscr{M}}(\Sigma){\restriction}_{\delta} := (\mathscr{M}(\Sigma){\restriction}_{\delta}, \sqsubseteq)/{\equiv}$ symbolizes the bounded lattice of (equivalence classes of) $\mathscr{M}$-concept descriptions the role depth of which is bounded by δ. Note that $\underline{\mathscr{M}}(\Sigma){\restriction}_{\delta}$ is indeed complete if the underlying signature Σ is finite, since then there are only finitely many $\mathcal{M}$-concept descriptions over Σ with a role depth of at most δ. Eventually, the *dual* $(\mathcal{M}(\Sigma))^{\partial}$ of the lattice $\mathcal{M}(\Sigma)$ is obtained by simply reversing the order relation, and an analogous notion applies to the lattice $\mathcal{M}(\Sigma)\!\restriction_{\delta}$.

6   Formal Concept Analysis

This section briefly introduces the standard notions of *Formal Concept Analysis* (abbr. *FCA*) [24]. A *formal context* $\mathbb{K} := (G, M, I)$ consists of a set G of *objects* (*Gegenstände* in German), a set M of *attributes* (*Merkmale* in German), and an *incidence relation* $I \subseteq G \times M$. For a pair $(g, m) \in I$, we say that g *has* m. The *derivation operators* of $\mathbb{K}$ are the mappings $\cdot^I \colon \wp(G) \to \wp(M)$ and $\cdot^I \colon \wp(M) \to \wp(G)$ such that for each object set $A \subseteq G$, the set A^I contains all attributes that are shared by all objects in A, and dually for each attribute set $B \subseteq M$, the set B^I contains all those objects that have all attributes from B. Formally, we define the derivation operators as follows.

$$A^I := \{\, m \in M \mid \forall\, g \in A \colon (g, m) \in I \,\} \quad \text{for object sets } A \subseteq G,$$

$$\text{and} \quad B^I := \{\, g \in G \mid \forall\, m \in B \colon (g, m) \in I \,\} \quad \text{for attribute sets } B \subseteq M.$$

For singleton sets, we may also use the abbreviations $g^I := \{g\}^I$ for all objects $g \in G$, as well as $m^I := \{m\}^I$ for all attributes $m \in M$.

It is well-known [24] that both derivation operators constitute a so-called *Galois connection* between the powersets $\wp(G)$ and $\wp(M)$, i.e., the following statements hold true for all subsets $A, A_1, A_2 \subseteq G$ and $B, B_1, B_2 \subseteq M$.

1. $A \subseteq B^I$ if, and only if, $B \subseteq A^I$ if, and only if, $A \times B \subseteq I$
2. $A \subseteq A^{II}$ 5. $B \subseteq B^{II}$
3. $A^I = A^{III}$ 6. $B^I = B^{III}$
4. $A_1 \subseteq A_2$ implies $A_2^I \subseteq A_1^I$ 7. $B_1 \subseteq B_2$ implies $B_2^I \subseteq B_1^I$

For obvious reasons, formal contexts can be represented as binary tables the rows of which are labeled with the objects, the columns of which are labeled with the attributes, and the occurrence of a cross $\times$ in the cell at row g and column m indicates that the object g has the attribute m.

An *intent* of $\mathbb{K}$ is an attribute set $B \subseteq M$ with $B = B^{II}$. The set of all intents of $\mathbb{K}$ is denoted by $\mathsf{Int}(\mathbb{K})$. An *implication* over M is an expression $X \to Y$ where $X, Y \subseteq M$. It is *valid* in $\mathbb{K}$, denoted as $\mathbb{K} \models X \to Y$, if $X^I \subseteq Y^I$, i.e., if each object of $\mathbb{K}$ that possesses all attributes in X also has all attributes in Y. An implication set $\mathscr{L}$ is *valid* in $\mathbb{K}$, denoted as $\mathbb{K} \models \mathscr{L}$, if all implications in $\mathscr{L}$ are valid in $\mathbb{K}$. Furthermore, the relation $\models$ is lifted to implication sets as follows: an implication set $\mathscr{L}$ *entails* an implication $X \to Y$, symbolized as $\mathscr{L} \models X \to Y$, if $X \to Y$ is valid in all formal contexts in which $\mathscr{L}$ is valid. More specifically, $\models$ is called the *semantic entailment relation*.

A *model* of $X \to Y$ is an attribute set $Z \subseteq M$ such that $X \subseteq Z$ implies $Y \subseteq Z$, and we shall then write $Z \models X \to Y$. Of course, then an implication $X \to Y$ is valid in $\mathbb{K}$ if, and only if, for each object $g \in G$, the *object intent* g^I is a model of $X \to Y$. It is furthermore straightforward to verify that the following statements are equivalent.

1. $X \to Y$ is valid in $\mathbb{K}$.
2. Each object intent of $\mathbb{K}$ is a model of $X \to Y$.
3. Each intent of $\mathbb{K}$ is a model of $X \to Y$.
4. $Y \subseteq X^{II}$.

The equivalence between the first and the last statement indicates that X^{II} is the largest consequence of X in $\mathbb{K}$, i.e., $X \to X^{II}$ is valid in $\mathbb{K}$, and for each strict superset $Z \supsetneq X^{II}$, the implication $X \to Z$ is not valid in $\mathbb{K}$.

Consider an implication set $\mathscr{L} \cup \{X \to Y\} \subseteq \mathsf{Imp}(M)$. A *model* of $\mathscr{L}$ is an attribute set which is a simultaneous model of each implication in $\mathscr{L}$. In particular, each model Z of $\mathscr{L}$ satisfies the following: for each implication $X \to Y \in \mathscr{L}$, $X \subseteq Z$ implies $Y \subseteq Z$, i.e., Z is a fixed point of the operator

$$Z \mapsto Z^{\mathscr{L}(1)} := Z \cup \bigcup \{Y \mid \exists X : X \to Y \in \mathscr{L} \text{ and } X \subseteq Z\}.$$

The smallest model $Z^{\mathscr{L}}$ of $\mathscr{L}$ that contains Z is obtained by successive exhaustive application of the operator $\cdot^{\mathscr{L}(1)}$, i.e., $Z^{\mathscr{L}} = \bigcup \{Z^{\mathscr{L}(n)} \mid n \geq 1\}$ where $Z^{\mathscr{L}(n+1)} := (Z^{\mathscr{L}(1)})^{\mathscr{L}(n)}$ for all $n \geq 1$. Additionally, the following statements are equivalent.

1. $\mathscr{L}$ entails $X \to Y$.
2. Each model of $\mathscr{L}$ is a model of $X \to Y$.
3. $X \to Y$ is valid in all those formal contexts with attribute set M in which $\mathscr{L}$ is valid.
4. $Y \subseteq X^{\mathscr{L}}$.

We then infer that $X^{\mathscr{L}}$ is the largest consequence of X with respect to the implication set $\mathscr{L}$, i.e., $\mathscr{L}$ entails $X \to X^{\mathscr{L}}$, and for all supersets $Y \supsetneq X^{\mathscr{L}}$, the implication $X \to Y$ does not follow from $\mathscr{L}$.

It was shown that entailment can also be decided *syntactically* by applying *deduction rules* to the implication set $\mathscr{L}$ without the requirement to consider all formal contexts in which $\mathscr{L}$ is valid, or all models of $\mathscr{L}$, respectively. Recall that an implication $X \to Y$ is *syntactically entailed* by an implication set $\mathscr{L}$, denoted by $\mathscr{L} \vdash X \to Y$, if $X \to Y$ can be constructed from $\mathscr{L}$ by the application of *inference axioms*, cf. [39, p. 47], which are described as follows.

(F1) *Reflexivity:* $\emptyset \vdash X \to X$
(F2) *Augmentation:* $\{X \to Y\} \vdash X \cup Z \to Y$
(F3) *Additivity:* $\{X \to Y, X \to Z\} \vdash X \to Y \cup Z$
(F4) *Projectivity:* $\{X \to Y \cup Z\} \vdash X \to Y$
(F5) *Transitivity:* $\{X \to Y, Y \to Z\} \vdash X \to Z$
(F6) *Pseudotransitivity:* $\{X \to Y, Y \cup Z \to W\} \vdash X \cup Z \to W$

In the inference axioms above the symbols X, Y, Z, and W denote arbitrary subsets of the considered set M of attributes. Formally, we define $\mathscr{L} \vdash X \to Y$ if there is a finite sequence of implications $X_0 \to Y_0, \ldots, X_n \to Y_n$ such that the following conditions hold.

1. For each $i \in \{0, \ldots, n\}$, there is a subset $\mathscr{L}_i \subseteq \mathscr{L} \cup \{X_0 \to Y_0, \ldots, X_{i-1} \to Y_{i-1}\}$ such that $\mathscr{L}_i \vdash X_i \to Y_i$ matches one of the Axioms F1–F6.
2. $X_n \to Y_n = X \to Y$.

Often, the Axioms F1, F2, and F6 are referred to as *Armstrong's axioms*. These three axioms constitute a *complete* and *independent* set of inference axioms for entailment, i.e., from it the other Axioms F3–F5 can be derived, and none of them is derivable from the others.

The semantic entailment and the syntactic entailment coincide, i.e., an implication $X \to Y$ is semantically entailed by an implication set $\mathscr{L}$ if, and only if, $\mathscr{L}$ syntactically entails $X \to Y$, cf. [39, Theorem 4.1 on Page 50] as well as [24, Proposition 21 on Page 81]. Consequently, we do not have to distinguish between both entailment relations $\models$ and $\vdash$ when it is up to decide whether an implication follows from a set of implications.

The data encoded in a formal context can be visualized as a *line diagram* of the corresponding *concept lattice*, which we shall shortly describe. A *formal concept* of a formal context $\mathbb{K} := (G, M, I)$ is a pair (A, B) consisting of a set $A \subseteq G$ of objects as well as a set $B \subseteq M$ of attributes such that $A^I = B$ and $B^I = A$. We then also refer to A as the *extent*, and to B as the *intent*, respectively, of (A, B). Another characterization of a formal concept is as follows: (A, B) is a formal concept of $\mathbb{K}$ if, and only if, $A \subseteq G$, $B \subseteq M$, and both A and B are maximal with respect to the property $A \times B \subseteq I$, i.e., for each strict superset $C \supsetneq A$, $C \times B \not\subseteq I$, and accordingly for each strict superset $D \supsetneq B$, $A \times D \not\subseteq I$. In the denotation of $\mathbb{K}$ as a cross table, those formal concepts are the maximal rectangles full of crosses (modulo reordering of rows and columns). Then, the set of all extents of $\mathbb{K}$ is symbolized as $\mathsf{Ext}(\mathbb{K})$, and the set of all formal concepts of $\mathbb{K}$ is denoted as $\mathfrak{B}(\mathbb{K})$, which is ordered by defining $(A, B) \leq (C, D)$ if, and only if, $A \subseteq C$. It was shown that this order always induces a complete lattice $\underline{\mathfrak{B}}(\mathbb{K}) := (\mathfrak{B}(\mathbb{K}), \leq, \wedge, \vee, \top, \bot)$, called the *concept lattice* of $\mathbb{K}$, cf. [24, 48], in which the infimum and the supremum operation satisfy the equations

$$\bigwedge \{(A_t, B_t) \mid t \in T\} = (\bigcap \{A_t \mid t \in T\}, (\bigcup \{B_t \mid t \in T\})^{II}),$$

and $$\bigvee \{(A_t, B_t) \mid t \in T\} = ((\bigcup \{A_t \mid t \in T\})^{II}, \bigcap \{B_t \mid t \in T\}),$$

and where $\top = (\emptyset^I, \emptyset^{II})$ is the greatest element, and where $\bot = (\emptyset^{II}, \emptyset^I)$ is the smallest element, respectively. The number of formal concepts can be exponential in the size of the formal context. Kuznetsov shows that determining this number is a #P-complete problem, cf. [34]. Furthermore, the problems of existence of a formal concept with restrictions on the size of the extent, intent, or both, respectively, are investigated in [34]—Kuznetsov demonstrates that the existence of a formal concept (A, B) such that $|A| = k$, $|B| = k$, or $|A| + |B| = k$, respectively, are NP-complete problems; the similar problems with $\geq$ are all in P; and the problems with $\leq$ are also in P, except the problem where $|A| + |B| \leq k$ is NP-complete.

Furthermore, the concept lattice of $\mathbb{K}$ can be nicely represented as a *line diagram* as follows: each formal concept is depicted as a vertex. Furthermore, there is an upward directed edge from each formal concept to its upper neighbors, i.e., to all those formal concepts which are greater with respect to $\leq$, but for which there is no other formal concept in between. The nodes are labeled as follows: an attribute $m \in M$ is an upper label of the *attribute concept* (m^I, m^{II}), and an object $g \in G$ is a lower label of the *object concept* (g^{II}, g^I). Then, the extent of the formal concept represented by a vertex consists of all objects which label vertices reachable by a downward directed path, and dually the intent is obtained by gathering all attribute labels of vertices reachable by an upward directed path.

Let $\mathbb{K} \models \mathscr{L}$. A *pseudo-intent* of a formal context $\mathbb{K}$ relative to an implication set $\mathscr{L}$ is an attribute set $P \subseteq M$ which is no intent of $\mathbb{K}$, but is a model of $\mathscr{L}$, and satisfies $Q^{II} \subseteq P$ for all pseudo-intents $Q \subsetneq P$. The set of all those pseudo-intents is symbolized by $\mathsf{PsInt}(\mathbb{K}, \mathscr{L})$. Then the implication set

$$\mathsf{Can}(\mathbb{K}, \mathscr{L}) := \{ P \to P^{II} \mid P \in \mathsf{PsInt}(\mathbb{K}, \mathscr{L}) \}$$

constitutes an *implication base* of $\mathbb{K}$ relative to $\mathscr{L}$, i.e., for each implication $X \to Y$ over M, the following equivalence is satisfied.

$$\mathbb{K} \models X \to Y \text{ if, and only if, } \mathsf{Can}(\mathbb{K}, \mathscr{L}) \cup \mathscr{L} \models X \to Y$$

$\mathsf{Can}(\mathbb{K}, \mathscr{L})$ is called the *canonical base* of $\mathbb{K}$ relative to $\mathscr{L}$. It can be shown that it is a *minimal* implication base of $\mathbb{K}$ relative to $\mathscr{L}$, i.e., there is no implication base of $\mathbb{K}$ relative to $\mathscr{L}$ with smaller cardinality. Further information is given in [21, 23, 25, 45]. The most prominent algorithm for computing the canonical base is certainly *NextClosure* developed by Ganter [21, 23]. Bazhanov and Obiedkov propose an optimized version of *NextClosure* in [8] which speeds up the computation of the lectically next closure, and furthermore they then perform some benchmarks to compare both versions. Additionally, they also utilize three different algorithms for computing closures with respect to implication sets, i.e., firstly the already presented and straightforward algorithm which computes the (least) fixed point of the operator $X \mapsto X^{\mathscr{L}(1)}$, see also [39], secondly the *LinClosure* algorithm [9], which computes $X^{\mathscr{L}}$ in linear time, and thirdly *Wild's Closure* algorithm [47], which is essentially an improved version of *LinClosure*. Please note that *LinClosure* is not always faster than computing the least fixed point of $X \mapsto X^{\mathscr{L}(1)}$, due to its initialization overhead. Furthermore, Obiedkov and Duquenne constitute an attribute-incremental algorithm for constructing the canonical base, cf. [42]. A parallel algorithm called *NextClosures* is also available [28, 32], and an implementation is provided in *Concept Explorer FX* [27]; its advantage is that its processing time scales almost inverse linear with respect to the number of available CPU cores.

There are some important complexity problems related to the pseudo-intents and canonical bases. Kuznetsov, and later together with Obiedkov, has proven in [35–37] that the number of pseudo-intents can be exponential in $|M|$ as well as in $|G| \cdot |M|$ or in $|I|$, and determining this number is #P-hard, furthermore that recognizing a pseudo-intent is in coNP, and that determining the number of non-pseudo-intents is #P-complete. Sertkaya and Distel demonstrated in [15, 17, 43, 44] that the number of intents can be exponential in the number of pseudo-intents, i.e., the set of pseudo-intents cannot be enumerated in output-polynomial time by utilizing one of the existing algorithms, which all enumerate the closure system of both intents and pseudo-intents, and that the lectically first pseudo-intent can be computed in polynomial time, but recognizing the first n pseudo-intents is coNP-complete. Consequently, the pseudo-intents of a given formal context cannot be enumerated in the lectic order with polynomial delay, unless P = NP. Enumeration of pseudo-intents (in an arbitrary order) was also investigated, but concrete complexity results are outstanding. Babin and Kuznetsov showed in [6, 7] that recognizing a pseudo-intent is coNP-complete, and furthermore that recognizing the lectically largest pseudo-intent is coNP-hard. Hence, computing pseudo-intents in the dual lectic order is also intractable, i.e., not possible with polynomial delay, unless P = NP. As a corollary Babin and Kuznetsov conclude that the maximal pseudo-intents cannot be enumerated with polynomial delay, unless P = NP. Further consequences which they found are, for example, that premises of minimal implication bases cannot be tractably recognized, since this problem is coNP-complete, and that there cannot be an algorithm that outputs a random pseudo-intent in polynomial time, unless NP = coNP.

Eventually, in case a given formal context is not complete in the sense that it does not contain enough objects to refute invalid implications, i.e., only contains some observed objects in the domain of interest, but one aims at exploring all valid implications over the given attribute set, a technique called *Attribute Exploration* can be utilized, which guides the user through the process of axiomatizing an implication base for the underlying domain in a way the number of questions posed to the user is minimal. For a sophisticated introduction as well as for theoretical and technical details, the interested reader is rather referred to [21–23, 31, 45]. A parallel variant of the *Attribute Exploration* also exists, cf. [28, 31], which is implemented in *Concept Explorer FX* [27].

For transferring and extending the results on canonical bases from *Formal Concept Analysis* to *Description Logics*, there are two key observations, namely that in the simple description logic $\mathscr{L}_0$, which only allows for $\top$ and $\sqcap$, there is a one-to-one correspondence between interpretations over the signature $(M, \emptyset)$ and formal contexts with attribute set M, and furthermore that implications over M can be represented as concept inclusions over $(M, \emptyset)$, and vice versa. In particular, an attribute subset $X \subseteq M$ then corresponds to the conjunction $\bigsqcap X$, and accordingly an implication $X \to Y$ corresponds to the CI $\bigsqcap X \sqsubseteq \bigsqcap Y$. These observations were successfully used in [2, 12, 16], among others. All of the aforementioned papers have in common that they provide a certain extension of the method for axiomatizing bases of implications from formal contexts. In particular, each of the methods makes heavy use of the *canonical base*. We will later elaborate on that, and provide results specifically tailored to our considered description logic $\mathcal{MH}$.

7 The Galois Connection of an Interpretation

In Sect. 6 we have seen that in *Formal Concept Analysis* the pair of the derivation operators $\cdot^I \colon \wp(G) \to \wp(M)$ and $\cdot^I \colon \wp(M) \to \wp(G)$ of a formal context $\mathbb{K} := (G, M, I)$ constitutes a Galois connection. In *Description Logics* however, for an interpretation $\mathcal{I} := (\Delta^{\mathcal{I}}, \cdot^{\mathcal{I}})$ we only have an *extension mapping* $\cdot^{\mathcal{I}} \colon \mathcal{M}(\Sigma) \to \wp(\Delta^{\mathcal{I}})$, which is defined recursively on the structure of concept descriptions, cf. Sect. 4. As a short repetition on Galois connections between posets, the interested reader is referred to [13, Definition 7.23] and [13, Lemma 7.26]. However, we will later formulate corresponding notions specifically tailored to our use case.

By definition the extension mapping $\cdot^{\mathcal{I}} \colon \mathcal{M}(\Sigma) \to \wp(\Delta^{\mathcal{I}})$ preserves finitary joins, i.e., we have that $(\bigsqcap\{ C_t \mid t \in T \})^{\mathcal{I}} = \bigcap\{ C_t^{\mathcal{I}} \mid t \in T \}$ for all finite families $\{ C_t \mid t \in T \}$ of $\mathcal{M}$-concept descriptions over Σ. When imposing a role-depth bound δ on the concept descriptions, then we know that there are only finitely many concept descriptions in case of a finite signature, and thus the extension mapping $\cdot^{\mathcal{I}} \colon \mathcal{M}(\Sigma){\restriction}_{\delta} \to \wp(\Delta^{\mathcal{I}})$ preserves arbitrary joins—then [13, 7.34] yields that there is another mapping $\wp(\Delta^{\mathcal{I}}) \to \mathcal{M}(\Sigma){\restriction}_{\delta}$, which together with $\cdot^{\mathcal{I}}$ constitutes a Galois connection, and in terms of lattice theory this mapping is called the *upper adjoint* of the extension mapping $\cdot^{\mathcal{I}}$. In [2, 12, 16] this upper adjoint is rather called *model-based most specific concept description mapping*, and in each of the references it was shown that the pair of this mapping together with the extension mapping forms a Galois connection. Furthermore, [13, 7.33] then states that this other mapping can be found as $X \mapsto \mathsf{Min}\{ C \in \mathcal{M}(\Sigma){\restriction}_{\delta} \mid X \subseteq C^{\mathcal{I}} \}$,[1] i.e., the mapping which assigns to each subset $X \subseteq \Delta^{\mathcal{I}}$ its role-depth-bounded model-based most specific concept description (or, to be formally correct, its equivalence class) which is characterized by the following definition.

---

[1] For a subset $X \subseteq P$ of a quasi-ordered set $(P, \leq)$, we use the expression $\mathsf{Min}(X)$ to denote the set of all those elements in X which are minimal with respect to $\leq$, i.e., $x \in \mathsf{Min}(X)$ if, and only if, $x \in X$ and there is no other element $y \in X$ such that $y \leq x$ and $y \not\equiv x$.

Definition 7.1 Let $\mathcal{I}$ be an interpretation over the signature $\Sigma = (N_C, N_R)$, and let $\delta \in \mathbb{N}$ be a role-depth bound. Then, for a subset $X \subseteq \Delta^{\mathcal{I}}$, a concept description $C \in \mathcal{M}(\Sigma) \upharpoonright_{\delta}$ is called *role-depth-bounded model-based most specific concept description* (abbr. *RMMSC*) of X in $\mathcal{I}$ with respect to δ if it satisfies the following conditions.

1. $\mathsf{rd}(C) \leq \delta$,
2. $X \subseteq C^{\mathcal{I}}$, and
3. for all $\mathcal{M}$-concept descriptions D over Σ with a role depth not exceeding δ, it holds true that $\emptyset \models C \sqsubseteq D$ if $X \subseteq D^{\mathcal{I}}$.

We shall denote the set of all RMMSCs in $\mathcal{I}$ w.r.t. δ by $\mathsf{Mmsc}(\mathcal{I}, \delta)$.

Firstly, all role-depth-bounded model-based most specific concept descriptions of X in $\mathcal{I}$ with respect to δ are equivalent, and a representative of the equivalence class is hence denoted as $X^{\mathcal{I}(\delta)}$. Secondly, we can easily convince us that $X^{\mathcal{I}(\delta)}$ always exists—provided that the underlying signature is finite. This is due to the fact that for a finite signature, only finitely many concept descriptions with a role depth of at most δ exist. Consequently, in order to construct $X^{\mathcal{I}(\delta)}$ we may just build the (finite) conjunction of all those concept descriptions the role depth of which does not exceed δ and the extension of which contains X as a subset. Of course, this does not yield a practical means for the construction of role-depth-bounded model-based most specific concept descriptions, but we will investigate an appropriate computation method later in Sect. 8.

Lemma 7.2 *Let $\mathcal{I}$ be an interpretation over the signature $\Sigma = (N_C, N_R)$, $\{ X_t \mid t \in T \}$ be a family of subsets $X_t \subseteq \Delta^{\mathcal{I}}$, and $\{ C_s \mid s \in S \}$ a family of concept descriptions $C_s \in \mathcal{M}(\Sigma)$. Then, the following statements hold.*

1. $\emptyset \models (\bigcup \{ X_t \mid t \in T \})^{\mathcal{I}(\delta)} \equiv \bigvee \{ X_t^{\mathcal{I}(\delta)} \mid t \in T \}$
2. $(\bigsqcap \{ C_s \mid s \in S \})^{\mathcal{I}} = \bigcap \{ C_s^{\mathcal{I}} \mid s \in S \}$

Proof

1. Let $\{ X_t \mid t \in T \}$ be a family of subsets $X_t \subseteq \Delta^{\mathcal{I}}$. Then we can show that $\bigvee \{ X_t^{\mathcal{I}(\delta)} \mid t \in T \}$ is indeed a role-depth-bounded model-based most specific concept description of $\bigcup \{ X_t \mid t \in T \}$. (It would also be possible to dually prove that $(\bigcup \{ X_t \mid t \in T \})^{\mathcal{I}(\delta)}$ is a least common subsumer of the concept descriptions $X_t^{\mathcal{I}(\delta)}$ for $t \in T$.)

   First, we prove that $\bigcup \{ X_t \mid t \in T \}$ is a subset of the extension $(\bigvee \{ X_t^{\mathcal{I}(\delta)} \mid t \in T \})^{\mathcal{I}}$. By definition, it holds that $X_t \subseteq X_t^{\mathcal{I}(\delta)\mathcal{I}}$ for all $t \in T$. Furthermore, every RMMSC $X_t^{\mathcal{I}(\delta)}$ is subsumed by the LCS $\bigvee \{ X_t^{\mathcal{I}(\delta)} \mid t \in T \}$. It then immediately follows that each X_t must be a subset of the extension $(\bigvee \{ X_t^{\mathcal{I}(\delta)} \mid t \in T \})^{\mathcal{I}}$.

   Second, we have to show that whenever C is a concept description the extension of which contains $\bigcup \{ X_t \mid t \in T \}$, then C subsumes $\bigvee \{ X_t^{\mathcal{I}(\delta)} \mid t \in T \}$ with respect to the empty TBox $\emptyset$. By definition of RMMSCs then we infer that each $X_t^{\mathcal{I}(\delta)}$ is subsumed by C, and hence by definition of LCS, $\bigvee \{ X_t^{\mathcal{I}(\delta)} \mid t \in T \}$ must be subsumed by C, too.
2. holds true by definition of the semantics of conjunctions.   □

Lemma 7.3 *Let $\mathcal{I}$ be an interpretation over the signature $\Sigma = (N_C, N_R)$, and $\delta \in \mathbb{N}$ be a role-depth bound. Then, the extension mapping $\cdot^{\mathcal{I}}$ and the MMSC-mapping $\cdot^{\mathcal{I}(\delta)}$ constitute a Galois connection between the powerset lattice of the domain $\Delta^{\mathcal{I}}$ and the dual of the concept description lattice $\underline{\mathcal{M}}(\Sigma)\!\restriction_{\delta}$.*

In particular, the following statements hold true for all subsets $X, Y \subseteq \Delta^{\mathcal{I}}$, and for all $\mathcal{M}$-concept descriptions C, D over Σ with a role-depth not exceeding δ.

1. $X \subseteq C^{\mathcal{I}}$ *if, and only if,* $\emptyset \models X^{\mathcal{I}(\delta)} \sqsubseteq C$
2. $X \subseteq X^{\mathcal{I}(\delta).\mathcal{I}}$
3. $\emptyset \models X^{\mathcal{I}(\delta)} \equiv X^{\mathcal{I}(\delta).\mathcal{I}.\mathcal{I}(\delta)}$
4. $X \subseteq Y$ *implies* $\emptyset \models X^{\mathcal{I}(\delta)} \sqsubseteq Y^{\mathcal{I}(\delta)}$
5. $\emptyset \models C \sqsupseteq C^{\mathcal{I}.\mathcal{I}(\delta)}$
6. $C^{\mathcal{I}} = C^{\mathcal{I}.\mathcal{I}(\delta).\mathcal{I}}$
7. $\emptyset \models C \sqsubseteq D$ *implies* $C^{\mathcal{I}} \subseteq D^{\mathcal{I}}$

Proof It suffices to prove the first statement, since the others are then obtained as consequences, cf. [13, Definition 7.23 and Lemma 7.26]. Hence, assume that $X \subseteq C^{\mathcal{I}}$. Then by Statement 3 of Definition 7.1 we conclude that $\emptyset \models X^{\mathcal{I}(\delta)} \sqsubseteq C$. Vice versa, if $X^{\mathcal{I}(\delta)}$ is subsumed by C with respect to the empty TBox $\emptyset$, then in particular it follows that $X^{\mathcal{I}(\delta).\mathcal{I}} \subseteq C^{\mathcal{I}}$. An application of Statement 2 of Definition 7.1 then yields $X \subseteq X^{\mathcal{I}(\delta).\mathcal{I}} \subseteq C^{\mathcal{I}}$. □

From the preceding lemma we conclude that the composition of the extension mapping and the MMSC mapping yields a closure operator in the dual of $\mathcal{M}\!\restriction_{\delta}$, and it furthermore holds true that the implications which are valid in $\cdot^{\mathcal{I}\,\mathcal{I}(\delta)}$ are exactly those concept inclusions which are valid in $\mathcal{I}$ and the subsumee and the subsumer of which have a role depth not exceeding δ. Furthermore, we infer that each implication base, of $\cdot^{\mathcal{I}\,\mathcal{I}(\delta)}$ is a base of CIs for $\mathcal{I}$ and δ. Further information on implications that are valid in closure operators can be found in [30, Sect. 3].

8 Computation of Role-Depth-Bounded Model-Based Most Specific Concept Descriptions

In this section we are going to develop a method for the computation of RMMSCs in $\mathcal{M}$. By definition of the $\mathcal{M}$-concept descriptions in Sect. 4, it follows that each such $\mathcal{M}$-concept description is essentially a conjunction of other $\mathcal{M}$-concept descriptions, i.e., for each $C \in \mathcal{M}(\Sigma)$, there is a finite set $\mathsf{Conj}(C) \subseteq \mathcal{M}(\Sigma)$ such that $C = \bigsqcap \mathsf{Conj}(C)^2$ is satisfied and $\mathsf{Conj}(C)$ does not contain any elements of the form $D \sqcap E$. We call the elements in $\mathsf{Conj}(C)$ the *top-level conjuncts* of C. Furthermore, we can distinguish between the different possible types of these top-level conjuncts, i.e., if $\mathcal{X} \subseteq \mathcal{M}(\Sigma)$, then $\mathsf{Conj}(C, \mathcal{X}) := \mathsf{Conj}(C) \cap \mathcal{X}$. If $\mathbf{A} \subseteq N_C$, $\mathbf{R} \subseteq N_R$, $\mathbf{N} \subseteq \mathbb{N}$, and $\mathbf{C} \subseteq \mathcal{M}(\Sigma)$, then define the following sets.

---

^{2}Please note that $\bigsqcap \emptyset = \top$.

$$\neg \mathbf{A} := \{\neg A \mid A \in \mathbf{A}\}$$

$$\forall \mathbf{R}.\mathbf{C} := \{\forall r.C \mid r \in \mathbf{R}, C \in \mathbf{C}\}$$

$$\exists \geq \mathbf{N}.\mathbf{R}.\mathbf{C} := \{\exists \geq n.r.C \mid n \in \mathbf{N}, r \in \mathbf{R}, C \in \mathbf{C}\}$$

$$\exists \leq \mathbf{N}.\mathbf{R} := \{\exists \leq n.r \mid n \in \mathbf{N}, r \in \mathbf{R}\}$$

$$\exists \mathbf{R}.\mathsf{Self} := \{\exists r.\mathsf{Self} \mid r \in \mathbf{R}\}$$

It is readily verified that then for every $\mathcal{M}$-concept description C,

$$
\begin{aligned}
\mathsf{Conj}(C) = \ & \mathsf{Conj}(C, \{\bot, \top\}) \\
& \cup \mathsf{Conj}(C, N_C) \\
& \cup \mathsf{Conj}(C, \neg N_C) \\
& \cup \mathsf{Conj}(C, \forall N_R.\mathcal{M}(\Sigma)) \\
& \cup \mathsf{Conj}(C, \exists \geq \mathbb{N}.N_R.\mathcal{M}(\Sigma)) \\
& \cup \mathsf{Conj}(C, \exists \leq \mathbb{N}.N_R) \\
& \cup \mathsf{Conj}(C, \exists N_R.\mathsf{Self}),
\end{aligned}
$$

i.e., C must be of the following form.

$$
\begin{aligned}
C = \ & \bigsqcap \mathsf{Conj}(C, \{\bot, \top\}) \\
& \sqcap \bigsqcap \mathsf{Conj}(C, N_C) \\
& \sqcap \bigsqcap \mathsf{Conj}(C, \neg N_C) \\
& \sqcap \bigsqcap \mathsf{Conj}(C, \forall N_R.\mathcal{M}(\Sigma)) \\
& \sqcap \bigsqcap \mathsf{Conj}(C, \exists \geq \mathbb{N}.N_R.\mathcal{M}(\Sigma)) \\
& \sqcap \bigsqcap \mathsf{Conj}(C, \exists \leq \mathbb{N}.N_R) \\
& \sqcap \bigsqcap \mathsf{Conj}(C, \exists N_R.\mathsf{Self})
\end{aligned}
$$

We conclude that for the construction of an RMMSC we have to investigate which conjuncts of the different types must occur in the RMMSC. In particular, we investigate a technique for the construction of an RMMSC $X^{\mathcal{I}(\delta)}$ of a subset $X \subseteq \Delta^{\mathcal{I}}$ within a given interpretation $\mathcal{I}$ and with respect to a pre-defined bound $\delta \in \mathbb{N}$ on the role depths. We start by considering the smallest bound $\delta = 0$. It is then readily verified that the RMMSC must have the form

$$X^{\mathscr{I}(0)} = \bigsqcap \mathsf{Conj}(X^{\mathscr{I}(0)}, \{\bot, \top\})$$

$$\sqcap \bigsqcap \mathsf{Conj}(X^{\mathscr{I}(0)}, N_C)$$

$$\sqcap \bigsqcap \mathsf{Conj}(X^{\mathscr{I}(0)}, \neg N_C),$$

where

$$\mathsf{Conj}(X^{\mathscr{I}(0)}, \{\bot, \top\}) = \{\top\} \cup \{\bot \mid X = \emptyset\},$$

$$\mathsf{Conj}(X^{\mathscr{I}(0)}, N_C) = \{A \mid A \in N_C \text{ and } X \subseteq A^{\mathscr{I}}\},$$

$$\text{and} \quad \mathsf{Conj}(X^{\mathscr{I}(0)}, \neg N_C) = \{\neg A \mid A \in N_C \text{ and } X \cap A^{\mathscr{I}} = \emptyset\}.$$

Now assume that $\delta > 0$. We have already argued that for a finite signature Σ, which we can always assume for practical cases, the RMMSC $X^{\mathscr{I}(\delta)}$ must exist, and furthermore must then be of the following form.

$$X^{\mathscr{I}(\delta)} = \bigsqcap \mathsf{Conj}(X^{\mathscr{I}(\delta)}, \{\bot, \top\})$$

$$\sqcap \bigsqcap \mathsf{Conj}(X^{\mathscr{I}(\delta)}, N_C)$$

$$\sqcap \bigsqcap \mathsf{Conj}(X^{\mathscr{I}(\delta)}, \neg N_C)$$

$$\sqcap \bigsqcap \mathsf{Conj}(X^{\mathscr{I}(\delta)}, \forall N_R. \mathscr{M}(\Sigma){\upharpoonright}_{\delta-1})$$

$$\sqcap \bigsqcap \mathsf{Conj}(X^{\mathscr{I}(\delta)}, \exists \geq \mathbb{N}. N_R. \mathscr{M}(\Sigma){\upharpoonright}_{\delta-1})$$

$$\sqcap \bigsqcap \mathsf{Conj}(X^{\mathscr{I}(\delta)}, \exists \leq \mathbb{N}. N_R)$$

$$\sqcap \bigsqcap \mathsf{Conj}(X^{\mathscr{I}(\delta)}, \exists N_R. \mathsf{Self})$$

For the first three parts, we can, of course, utilize the results from the case $\delta = 0$. Furthermore, we can immediately see that

$$\mathsf{Conj}(X^{\mathscr{I}(\delta)}, \exists N_R. \mathsf{Self}) = \{\exists r. \mathsf{Self} \mid r \in N_R \text{ and } \forall x \in X: (x, x) \in r^{\mathscr{I}}\}.$$

For analyzing the remaining parts, we repeat the definitions of extensions of some of the corresponding $\mathscr{M}$-concept descriptions as follows.

$$(\forall r. C)^{\mathscr{I}} = \{d \in \Delta^{\mathscr{I}} \mid \forall e \in \Delta^{\mathscr{I}}: (d, e) \in r^{\mathscr{I}} \text{ implies } e \in C^{\mathscr{I}}\}$$

$$= \{d \in \Delta^{\mathscr{I}} \mid \{e \in \Delta^{\mathscr{I}} \mid (d, e) \in r^{\mathscr{I}}\} \subseteq C^{\mathscr{I}}\}$$

$$(\exists \geq n.r.\,C)^{\mathscr{I}} = \{d \in \Delta^{\mathscr{I}} \mid |\{e \in \Delta^{\mathscr{I}} \mid (d,e) \in r^{\mathscr{I}} \text{ and } e \in C^{\mathscr{I}}\}| \geq n\}$$

$$= \{d \in \Delta^{\mathscr{I}} \mid |\{e \in \Delta^{\mathscr{I}} \mid (d,e) \in r^{\mathscr{I}}\} \cap C^{\mathscr{I}}| \geq n\}$$

$$(\exists \leq n.r)^{\mathscr{I}} = \{d \in \Delta^{\mathscr{I}} \mid |\{e \in \Delta^{\mathscr{I}} \mid (d,e) \in r^{\mathscr{I}}\}| \leq n\}$$

If we denote the set of all r-successors of an element $d \in \Delta^{\mathscr{I}}$ by $\mathsf{suc}_{\mathscr{I}}(d,r)$, i.e., if we set $\mathsf{suc}_{\mathscr{I}}(d,r) := \{e \in \Delta^{\mathscr{I}} \mid (d,e) \in r^{\mathscr{I}}\}$, then we can rewrite the equations given above as follows.

$$(\forall r.\,C)^{\mathscr{I}} = \{d \in \Delta^{\mathscr{I}} \mid \mathsf{suc}_{\mathscr{I}}(d,r) \subseteq C^{\mathscr{I}}\}$$

$$(\exists \geq n.r.\,C)^{\mathscr{I}} = \{d \in \Delta^{\mathscr{I}} \mid |\mathsf{suc}_{\mathscr{I}}(d,r) \cap C^{\mathscr{I}}| \geq n\}$$

$$(\exists \leq n.r)^{\mathscr{I}} = \{d \in \Delta^{\mathscr{I}} \mid |\mathsf{suc}_{\mathscr{I}}(d,r)| \leq n\}$$

Consequently, when lifting the equations from a characterization of elements of the extensions to subsets of the extensions, we get the following equivalences.

$$X \subseteq (\forall r.\,C)^{\mathscr{I}} \text{ if, and only if, } \forall x \in X{:}\, x \in (\forall r.\,C)^{\mathscr{I}}$$

$$\text{if, and only if, } \forall x \in X{:}\, \mathsf{suc}_{\mathscr{I}}(x,r) \subseteq C^{\mathscr{I}}$$

$$X \subseteq (\exists \geq n.r.\,C)^{\mathscr{I}} \text{ if, and only if, } \forall x \in X{:}\, x \in (\exists \geq n.r.\,C)^{\mathscr{I}}$$

$$\text{if, and only if, } \forall x \in X{:}\, |\mathsf{suc}_{\mathscr{I}}(x,r) \cap C^{\mathscr{I}}| \geq n$$

$$X \subseteq (\exists \leq n.r)^{\mathscr{I}} \text{ if, and only if, } \forall x \in X{:}\, x \in (\exists \leq n.r)^{\mathscr{I}}$$

$$\text{if, and only if, } \forall x \in X{:}\, |\mathsf{suc}_{\mathscr{I}}(x,r)| \leq n$$

Further define

$$\mathsf{CSuc}(X, \forall r) := \{C \in \mathscr{M}(\Sigma) \mid \forall x \in X{:}\, \mathsf{suc}_{\mathscr{I}}(x,r) \subseteq C^{\mathscr{I}}\},$$

$$\mathsf{CSuc}(X, \exists \geq n.r) := \{C \in \mathscr{M}(\Sigma) \mid \forall x \in X{:}\, |\mathsf{suc}_{\mathscr{I}}(x,r) \cap C^{\mathscr{I}}| \geq n\},$$

$$\text{and} \quad n(X,r) := \mathsf{max}\{\,|\mathsf{suc}_{\mathscr{I}}(x,r)| \mid x \in X\,\},$$

i.e., $n(x,r)$ denotes the number of r-successors of x in $\mathscr{I}$, and $n(X,r)$ is the smallest n such that $X \subseteq (\exists \leq n.r)^{\mathscr{I}}$. Then, of course it holds true that

$$X \subseteq (\forall r.\,C)^{\mathscr{I}} \text{ if, and only if, } C \in \mathsf{CSuc}(X, \forall r),$$

$$X \subseteq (\exists \geq n.r.\,C)^{\mathscr{I}} \text{ if, and only if, } C \in \mathsf{CSuc}(X, \exists \geq n.r),$$

$$\text{and} \quad X \subseteq (\exists \leq n.r)^{\mathscr{I}} \text{ if, and only if, } n \geq n(X,r).$$

We can then collect all subsets of the interpretation's domain the extension of which serves as a filler for the appropriate constructors, and in particular we set

$$\mathsf{Suc}_{\mathscr{I}}(X, \forall r) := \{ Y \subseteq \Delta^{\mathscr{I}} \mid \forall x \in X \colon \mathsf{suc}_{\mathscr{I}}(x, r) \subseteq Y \},$$

and $\quad \mathsf{Suc}_{\mathscr{I}}(X, \exists \geq n.r) := \{ Y \subseteq \Delta^{\mathscr{I}} \mid \forall x \in X \colon |\mathsf{suc}_{\mathscr{I}}(x, r) \cap Y| \geq n \}.$

Obviously, then

$$X \subseteq (\forall r. Y^{\mathscr{I}(\delta-1)})^{\mathscr{I}} \qquad \text{for all } Y \in \mathsf{Suc}_{\mathscr{I}}(X, \forall r),$$

and $\quad X \subseteq (\exists \geq n.r. Y^{\mathscr{I}(\delta-1)})^{\mathscr{I}} \text{ for all } Y \in \mathsf{Suc}_{\mathscr{I}}(X, \exists \geq n.r),$

and applying Statement 1 of Lemma 7.3 yields that

$$\emptyset \models X^{\mathscr{I}(\delta)} \sqsubseteq \forall r. Y^{\mathscr{I}(\delta-1)} \qquad \text{for all } Y \in \mathsf{Suc}_{\mathscr{I}}(X, \forall r),$$

$$\emptyset \models X^{\mathscr{I}(\delta)} \sqsubseteq \exists \geq n.r. Y^{\mathscr{I}(\delta-1)} \text{ for all } Y \in \mathsf{Suc}_{\mathscr{I}}(X, \exists \geq n.r),$$

and $\quad \emptyset \models X^{\mathscr{I}(\delta)} \sqsubseteq \exists \leq n.r \qquad \text{for all } n \geq n(X, r).$

The connection between the sets $\mathsf{CSuc}(\dots)$ and $\mathsf{Suc}(\dots)$ is as follows.

1. For all $C \in \mathsf{CSuc}(X, \mho r)$ it holds true that $C^{\mathscr{I}} \in \mathsf{Suc}(X, \mho r)$.
2. For all $Y \in \mathsf{Suc}(X, \mho r)$ it holds true that $Y^{\mathscr{I}(\delta-1)} \in \mathsf{CSuc}(X, \mho r)$.

Continuing the way towards a construction of the RMMSC of a subset $X \subseteq \Delta^{\mathscr{I}}$, we can see that it must satisfy the following subsumption.

$$\emptyset \models X^{\mathscr{I}(\delta)} \sqsubseteq \prod \{ A \mid A \in N_C \text{ and } X \subseteq A^{\mathscr{I}} \}$$

$$\sqcap \prod \{ \neg A \mid A \in N_C \text{ and } X \subseteq (\neg A)^{\mathscr{I}} \}$$

$$\sqcap \prod \{ \forall r. C \mid r \in N_R, \ C \in \mathscr{M}(\Sigma) \restriction_{\delta-1}, \text{ and } X \subseteq (\forall r. C)^{\mathscr{I}} \}$$

$$\sqcap \prod \left\{ \exists \geq n.r. C \ \middle| \ \begin{matrix} n \in \mathbb{N}, \ r \in N_R, \ C \in \mathscr{M}(\Sigma) \restriction_{\delta-1}, \\ \text{and } X \subseteq (\exists \geq n.r. C)^{\mathscr{I}} \end{matrix} \right\}$$

$$\sqcap \prod \{ \exists \leq n.r \mid n \in \mathbb{N}, \ r \in N_R, \text{ and } X \subseteq (\exists \leq n.r)^{\mathscr{I}} \}$$

$$\sqcap \prod \{ \exists r. \mathsf{Self} \mid r \in N_R, \text{ and } X \subseteq (\exists r. \mathsf{Self})^{\mathscr{I}} \}$$

$$\equiv \prod \{ A \mid A \in N_C \text{ and } X \subseteq A^{\mathscr{I}} \}$$

$$\sqcap \prod \{ \neg A \mid A \in N_C \text{ and } X \cap A^{\mathscr{I}} = \emptyset \}$$

$$\sqcap \prod \{ \forall r. C \mid r \in N_R, \text{ and } C \in \mathsf{CSuc}(X, \forall r) \cap \mathscr{M}(\Sigma) \restriction_{\delta-1} \}$$

$$\sqcap \bigsqcap \left\{ \exists \geq n.r.\, C \;\middle|\; \begin{array}{l} n \in \mathbb{N},\ r \in N_R, \\ \text{and } C \in \mathsf{CSuc}(X, \exists \geq n.r) \cap \mathscr{M}(\Sigma){\upharpoonright}_{\delta-1} \end{array} \right\}$$

$$\sqcap \bigsqcap \{ \exists \leq n.r \mid n \in \mathbb{N},\ r \in N_R,\ \text{and } n \geq n(X,r) \}$$

$$\sqcap \bigsqcap \{ \exists r.\, \mathsf{Self} \mid r \in N_R,\ \text{and } X \subseteq (\exists r.\, \mathsf{Self})^{\mathscr{I}} \}$$

It is easy to see that for the construction of the RMMSC it suffices to consider the minimal successors, and hence we explicitly define them as follows.

$$\mathsf{suc}_{\mathscr{I}}(X, r) := \bigcup \{ \mathsf{suc}_{\mathscr{I}}(x, r) \mid x \in X \}$$

$$= \{ y \in \Delta^{\mathscr{I}} \mid \exists x \in X \colon (x, y) \in r^{\mathscr{I}} \}$$

$$\mathsf{MinSuc}_{\mathscr{I}}(X, \forall r) := \mathsf{Min}(\mathsf{Suc}_{\mathscr{I}}(X, \forall r))$$

$$= \{ \mathsf{suc}_{\mathscr{I}}(X, r) \}$$

$$\mathsf{MinSuc}_{\mathscr{I}}(X, \exists \geq n.r) := \mathsf{Min}(\mathsf{Suc}_{\mathscr{I}}(X, \exists \geq n.r))$$

$$= \mathsf{Min}\{ Y \subseteq \mathsf{suc}_{\mathscr{I}}(X, r) \mid \forall x \in X \colon |\mathsf{suc}_{\mathscr{I}}(x, r) \cap Y| \geq n \}$$

Definition 8.1 Let $\mathscr{I}$ be a finite interpretation over a finite signature $\Sigma :=$ (N_C, N_R), $X \subseteq \Delta^{\mathscr{I}}$ with $X \neq \emptyset$ be a subset of the domain, and $\delta \in \mathbb{N}$ be a role-depth bound. Then, the *syntactic RMMSC* of X in $\mathscr{I}$ with respect to δ is the concept description $\mathsf{mmsc}(X, \mathscr{I}, \delta)$ which is defined by induction on the role depth as follows.

$$\mathsf{mmsc}(X, \mathscr{I}, 0) := \bigsqcap \{ A \mid A \in N_C \text{ and } X \subseteq A^{\mathscr{I}} \}$$

$$\sqcap \bigsqcap \{ \neg A \mid A \in N_C \text{ and } X \cap A^{\mathscr{I}} = \emptyset \}$$

$$\mathsf{mmsc}(X, \mathscr{I}, \delta) := \mathsf{mmsc}(X, \mathscr{I}, 0)$$

$$\sqcap \bigsqcap \left\{ \forall r.\, \mathsf{mmsc}(Y, \mathscr{I}, \delta - 1) \;\middle|\; \begin{array}{l} r \in N_R \\ \text{and } Y \in \mathsf{MinSuc}_{\mathscr{I}}(X, \forall r) \end{array} \right\}$$

$$\sqcap \bigsqcap \left\{ \exists \geq n.r.\, \mathsf{mmsc}(Y, \mathscr{I}, \delta - 1) \;\middle|\; \begin{array}{l} n \in \mathbb{N}_+,\ r \in N_R,\ \text{and} \\ Y \in \mathsf{MinSuc}_{\mathscr{I}}(X, \exists \geq n.r) \end{array} \right\}$$

$$\sqcap \bigsqcap \{ \exists \leq n(X, r).r \mid r \in N_R \}$$

$$\sqcap \bigsqcap \{ \exists r.\, \mathsf{Self} \mid r \in N_R \text{ and } \{ (x, x) \mid x \in X \} \subseteq r^{\mathscr{I}} \}$$

Furthermore, we define $\mathsf{mmsc}(\emptyset, \mathscr{I}, \delta) := \bot$ for all $\delta \in \mathbb{N}$.

Lemma 8.2 *Let $C_1, \ldots, C_m$ and $D_1, \ldots, D_n$ be $\mathcal{M}$-concept descriptions over the signature $\Sigma := (N_C, N_R)$. Then $\emptyset \models \bigsqcap \{ C_i \mid i \in \{1, \ldots, m\} \} \sqsubseteq \bigsqcap \{ D_j \mid j \in \{1, \ldots, n\} \}$ if for each $j \in \{1, \ldots, n\}$, there is an $i \in \{1, \ldots, m\}$ such that $\emptyset \models C_i \sqsubseteq D_j$.*

Proof Obviously, it holds true that $\emptyset \models \bigsqcap \{ C_i \mid i \in \{1, \ldots, m\} \} \sqsubseteq C_i$ for all indices $i \in \{1, \ldots, m\}$. We conclude that for each $j \in \{1, \ldots, n\}$, the subsumption $\emptyset \models \bigsqcap \{ C_i \mid i \in \{1, \ldots, m\} \} \sqsubseteq D_j$ is satisfied, and thus $\emptyset \models \bigsqcap \{ C_i \mid i \in \{1, \ldots, m\} \} \sqsubseteq \bigsqcap \{ D_j \mid j \in \{1, \ldots, n\} \}$. □

Theorem 8.3 *Let $\mathcal{I}$ be a finite interpretation over a finite signature $\Sigma := (N_C, N_R)$, $X \subseteq \Delta^{\mathcal{I}}$ a subset of the domain, and $\delta \in \mathbb{N}$ a role-depth bound. Then, the concept description $\mathsf{mmsc}(X, \mathcal{I}, \delta)$ is the role-depth-bounded model-based most-specific concept description of X in $\mathcal{I}$ with respect to δ, i.e., $\emptyset \models X^{\mathcal{I}(\delta)} \equiv \mathsf{mmsc}(X, \mathcal{I}, \delta)$.*

Proof The case $X = \emptyset$ is obvious. Hence, consider a non-empty subset $X \subseteq \Delta^{\mathcal{I}}$. It is easy to see that for a finite interpretation $\mathcal{I}$, it always holds true that $\mathsf{MinSuc}_{\mathcal{I}}(X, \exists \geq n. r) = \emptyset$ for all numbers $n > |\Delta^{\mathcal{I}}|$ and all role names $r \in N_R$. Consequently $\mathsf{mmsc}(X, \mathcal{I}, \delta)$ consists of finitely many conjunctions, and thus is indeed a well-defined $\mathcal{M}$-concept description.

We now show the three properties of Definition 7.1 by simultaneous induction on the role-depth bound δ.

($\delta = 0$) 1. Since concept names and their negations possess a role depth of 0, it obviously follows that $\mathsf{mmsc}(X, \mathcal{I}, 0)$ must have a role-depth of 0, too.

2. Since for each concept name $A \in N_C$ occurring in $\mathsf{mmsc}(X, \mathcal{I}, 0)$, it is true that $X \subseteq A^{\mathcal{I}}$, and furthermore for each primitive negation $\neg A$ for an $A \in N_C$ which is a top-level conjunct in $\mathsf{mmsc}(X, \mathcal{I}, 0)$, we have that $X \subseteq \Delta^{\mathcal{I}} \setminus A^{\mathcal{I}}$, we can easily conclude that $X \subseteq \mathsf{mmsc}(X, \mathcal{I}, 0)^{\mathcal{I}}$.

3. Assume that D is an $\mathcal{M}$-concept description over Σ with a role depth of 0, i.e., D consists only of a conjunction of concept names and primitive negations, and let $X \subseteq D^{\mathcal{I}}$. Then, for concept name $A \in N_C$ occurring in D, it certainly holds that $X \subseteq A^{\mathcal{I}}$, and hence A is a top-level conjunct in $\mathsf{mmsc}(X, \mathcal{I}, 0)$, too. Analogously, for a primitive negation $\neg A$ in D, we know that $X \subseteq (\neg A)^{\mathcal{I}}$ must be satisfied, and so also $\neg A$ is contained in the top-level conjunction of $\mathsf{mmsc}(X, \mathcal{I}, 0)$. We just showed that each conjunct in D also occurs in $\mathsf{mmsc}(X, \mathcal{I}, 0)$, and hence $\emptyset \models \mathsf{mmsc}(X, \mathcal{I}, 0) \sqsubseteq D$.

($\delta > 0$) 1. Note that $\mathsf{rd}(\mathsf{mmsc}(X, \mathcal{I}, \delta)) = 1 + \max \{ \mathsf{rd}(\mathsf{mmsc}(Y, \mathcal{I}, \delta - 1)) \mid Y \in \mathsf{MinSuc}(X, \mho r), \mho \in \{\forall\} \cup \{ \geq n. \mid n \in \mathbb{N}_+ \} \}$ for $\delta > 0$. By induction hypothesis, $\mathsf{rd}(\mathsf{mmsc}(Y, \mathcal{I}, \delta - 1)) \leq \delta - 1$, and hence it follows that $\mathsf{rd}(\mathsf{mmsc}(X, \mathcal{I}, \delta)) \leq \delta$.

2. Let $\delta > 0$, and consider a top-level conjunct $\mho r. \mathsf{mmsc}(Y, \mathcal{I}, \delta - 1)$ occurring in $\mathsf{mmsc}(X, \mathcal{I}, \delta)$, i.e., $Y \in \mathsf{MinSuc}_{\mathcal{I}}(X, \mho r)$. By induction hypothesis, Y is a subset of $\mathsf{mmsc}(Y, \mathcal{I}, \delta - 1)^{\mathcal{I}}$. We continue with a case distinction on the quantifier $\mho$.

($\mho = \geq n$) By definition of the successor sets, it holds true that all elements in Y are r-successors of some element in X, since $Y \subseteq \mathsf{suc}_{\mathscr{I}}(X, r)$. Furthermore, Y satisfies the condition that for each element $x \in X$, the cardinality of the intersection $\mathsf{suc}_{\mathscr{I}}(x, r) \cap Y$ is at least n, i.e., each element $x \in X$ has n or more r-successors in Y. Consequently, $X \subseteq (\exists \geq n.\, r.\, \mathsf{mmsc}(Y, \mathscr{I}, \delta - 1))^{\mathscr{I}}$.

($\mho = \forall$) In this case, we have that $Y = \mathsf{suc}_{\mathscr{I}}(X, r)$. Consider an arbitrary $x \in X$. If $y \in \Delta^{\mathscr{I}}$ and $(x, y) \in r^{\mathscr{I}}$, then $y \in Y$, and so $x \in (\forall r.\, \mathsf{mmsc}(Y, \mathscr{I}, \delta - 1))^{\mathscr{I}}$.

3. Consider $\delta > 0$, and let E be a conjunct on the top-level of D. Of course, it then holds true that $X \subseteq E^{\mathscr{I}}$. We proceed with a case distinction on E, and prove that there is always a top-level conjunct in $\mathsf{mmsc}(X, \mathscr{I}, \delta)$ which is subsumed by E with respect to the empty TBox $\emptyset$. As a consequence then Lemma 8.2 yields that $\emptyset \models \mathsf{mmsc}(X, \mathscr{I}, \delta) \sqsubseteq D$.

($E = \forall r.\, F$) Since $X \subseteq (\forall r.\, F)^{\mathscr{I}}$, we infer that each r-successor of each element in X is in the extension $F^{\mathscr{I}}$, i.e.,

$$\forall x \in X \, \forall y \in \Delta^{\mathscr{I}} : (x, y) \in r^{\mathscr{I}} \text{ implies } y \in F^{\mathscr{I}}.$$

As the set $\mathsf{suc}_{\mathscr{I}}(X, r)$ contains all r-successors of any element in X and no additional elements, we conclude that $\mathsf{suc}_{\mathscr{I}}(X, r) \subseteq F^{\mathscr{I}}$. Applying Statement 1 of Lemma 7.3 yields $\emptyset \models (\mathsf{suc}_{\mathscr{I}}(X, r))^{\mathscr{I}(\delta-1)} \sqsubseteq F$. An application of the induction hypothesis implies that $\emptyset \models (\mathsf{suc}_{\mathscr{I}}(X, r))^{\mathscr{I}(\delta-1)} \equiv \mathsf{mmsc}(\mathsf{suc}_{\mathscr{I}}(X, r), \mathscr{I}, \delta - 1)$. Eventually, it follows that

$$\emptyset \models \forall r.\, \mathsf{mmsc}(\mathsf{suc}_{\mathscr{I}}(X, r), \mathscr{I}, \delta - 1) \sqsubseteq \forall r.\, F.$$

($E = \exists \geq n.\, r.\, F$) By assumption, we have that $X \subseteq (\exists \geq n.\, r.\, F)^{\mathscr{I}}$, i.e., every element $x \in X$ has n or more r-successors which are in the extension of F. Thus, $|\mathsf{suc}_{\mathscr{I}}(x, r) \cap F^{\mathscr{I}}| \geq n$ for all $x \in X$, and consequently there is a set $Y \in \mathsf{MinSuc}_{\mathscr{I}}(X, \exists \geq n.\, r)$ such that $Y \subseteq F^{\mathscr{I}}$. By applying Statement 1 of Lemma 7.3 we conclude that $\emptyset \models Y^{\mathscr{I}(\delta-1)} \sqsubseteq F$, and since the induction hypothesis yields that $\emptyset \models Y^{\mathscr{I}(\delta-1)} \equiv \mathsf{mmsc}(Y, \mathscr{I}, \delta - 1)$, it eventually follows that $\emptyset \models \exists \geq n.\, r.\, \mathsf{mmsc}(Y, \mathscr{I}, \delta - 1) \sqsubseteq \exists \geq n.\, r.\, F$ where the subsumee is a top-level conjunct in $\mathsf{mmsc}(X, \mathscr{I}, \delta)$.

($E = \exists \leq n.\, r$) The set inclusion $X \subseteq (\exists \leq n.\, r)^{\mathscr{I}}$ yields that for every element $x \in X$, the number of r-successors of x does not exceed n. It is readily verified that then $n(X, r) \leq n$, and thus $\emptyset \models \exists \leq n(X, r).\, r \sqsubseteq \exists \leq n.\, r$. Of course, $\exists \leq n(X, r).\, r$ is contained as a top-level conjunct in $\mathsf{mmsc}(X, \mathscr{I}, \delta)$.

($E = \exists r.\, \mathsf{Self}$) From $X \subseteq (\exists r.\, \mathsf{Self})^{\mathscr{I}}$ it follows that each element $x \in X$ is an r-successor of itself, i.e., $\{\, (x, x) \mid x \in X \,\} \subseteq r^{\mathscr{I}}$. By definition, $\mathsf{mmsc}(X, \mathscr{I}, \delta)$ then also contains $\exists r.\, \mathsf{Self}$ as a top-level conjunct. $\square$

9   Concept Lattices of Interpretations

Let $\mathscr{I}$ be an interpretation over $\Sigma := (N_C, N_R)$, and assume that $\delta \in \mathbb{N}$ is a role depth bound. A *formal concept* of $\mathscr{I}$ with respect to the role depth bound δ is a pair $(X, [C]_{\equiv})$ such that its *extent* X is a subset of $\Delta^{\mathscr{I}}$, its *intent* $[C]_{\equiv}$ is an equivalence class of $\mathscr{M}$-concept descriptions over Σ, and $X^{\mathscr{I}(\delta)} = [C]_{\equiv}$ as well as $C^{\mathscr{I}} = X$ are satisfied. For the sake of simplicity, we denote the formal concept $(X, [C]_{\equiv})$ simply as (X, C). Then we may furthermore define an ordering of formal concepts by $(X, C) \leq (Y, D)$ if $X \subseteq Y$. In case $(X, C) \leq (Y, D)$ we say that (X, C) is a *subconcept* of (Y, D), and vice versa that (Y, D) is a *superconcept* of (X, C). Using the Galois properties from Lemma 7.3, it is easy to prove that $(X, C) \leq (Y, D)$ if, and only if, $\emptyset \models C \sqsubseteq D$. The set of all formal concepts of $\mathscr{I}$ w.r.t. δ is denoted by $\mathfrak{B}(\mathscr{I}, \delta)$, and the set of all extents is symbolized as $\mathsf{Ext}(\mathscr{I}, \delta)$.

Lemma 9.1 *Let $\mathscr{I}$ be a finite interpretation over the signature Σ, and $\delta \in \mathbb{N}$ a role-depth bound.*

1. *For all formal concepts (X, C) and (Y, D) of $\mathscr{I}$ w.r.t. δ, it is true that*

$$(X, C) \leq (Y, D) \text{ if, and only if, } X \subseteq Y \text{ if, and only if, } \emptyset \models C \sqsubseteq D.$$

2. *The relation $\leq$ is an order on $\mathfrak{B}(\mathscr{I}, \delta)$.*

Proof

1. The first equivalence holds by definition. Assume that X is a subset of Y, then from Statement 4 of Lemma 7.3 it follows that $\emptyset \models X^{\mathscr{I}(\delta)} \sqsubseteq Y^{\mathscr{I}(\delta)}$. Finally, since (X, C) and (Y, D) are description concepts we conclude $\emptyset \models C \equiv X^{\mathscr{I}(\delta)} \sqsubseteq Y^{\mathscr{I}(\delta)} \equiv D$. The other direction can be shown analogously, as also the extension mapping is monotonous, cf. Statement 7 of Lemma 7.3.
2. It is well-known that the subset inclusion is an order relation, hence also $\leq$ must be reflexive and transitive. □

Furthermore, $\mathfrak{B}(\mathscr{I}, \delta)$ is in fact a lattice, in which the *infimum* and the *supremum* of a finite family $\{ (X_t, C_t) \mid t \in T \}$ of formal concepts satisfy the following equations.

$$\bigwedge \{ (X_t, C_t) \mid t \in T \} = (\bigcap \{ X_t \mid t \in T \}, (\bigsqcap \{ C_t \mid t \in T \})^{\mathscr{I}\mathscr{I}(\delta)})$$

$$\bigvee \{ (X_t, C_t) \mid t \in T \} = ((\bigcup \{ X_t \mid t \in T \})^{\mathscr{I}(\delta)\mathscr{I}}, \bigvee \{ C_t \mid t \in T \})$$

The lattice is *bounded* by the *smallest* formal concept $(\emptyset, \bot)$, and by the *greatest* formal concept $(\Delta^{\mathscr{I}}, (\Delta^{\mathscr{I}})^{\mathscr{I}})$. We denote this lattice by $\underline{\mathfrak{B}}(\mathscr{I}, \delta) := (\mathfrak{B}(\mathscr{I}, \delta), \leq)$. Note that in case of finiteness of the interpretation $\mathscr{I}$, the concept lattice is *complete*.

10   Induced Formal Contexts

In this section we are going to consider the notion of *induced formal contexts*, which has first been defined and utilized by Baader and Distel [2, 16], and later also by Borchmann [11], for the description logic $\mathcal{E}\mathcal{L}^{\perp}_{\text{gfp}}$. Similar results were found by Borchmann, Distel, and Kriegel, cf. [12], for the description logic $\mathcal{E}\mathcal{L}^{\perp}$ where the role depth of the considered concept descriptions is restricted. In the sequel of this section, we extend the previous definitions and results to the more expressive description logic $\mathcal{M}$.

Consider a set $\mathcal{C}$ of $\mathcal{M}$-concept descriptions over the signature $\Sigma := (N_C, N_R)$. Then, we define a *projection* $\pi_{\mathcal{C}}$ with respect to $\mathcal{C}$ as follows.

$$\pi_{\mathcal{C}}: \mathcal{M}(\Sigma) \to \wp(\mathcal{C})$$

$$C \mapsto \{D \in \mathcal{C} \mid \emptyset \models C \sqsubseteq D\}$$

Furthermore, we say that an $\mathcal{M}$-concept description C over Σ is *expressible in terms of* $\mathcal{C}$ if there is a subset $\mathcal{X} \subseteq \mathcal{C}$ such that $\emptyset \models C \equiv \bigsqcap \mathcal{X}$. It turns out that the projection $\pi_{\mathcal{C}}$ is a counterpart for the conjunction $\bigsqcap$ such that their pair constitutes a Galois connection between the lattice $\mathcal{M}(\Sigma)$ and the powerset $\wp(\mathcal{C})$, i.e., the statements in the following lemma hold true.

Lemma 10.1 *Let $\mathcal{C}$ be a set of $\mathcal{M}$-concept descriptions over Σ. Then for all subsets $\mathcal{X}, \mathcal{Y} \subseteq \mathcal{C}$ and all concept descriptions $C, D \in \mathcal{M}(\Sigma)$, the following statements are valid.*

1. $\mathcal{X} \subseteq \pi_{\mathcal{C}}(C)$ *if, and only if,* $\emptyset \models \bigsqcap \mathcal{X} \sqsupseteq C$
2. $\mathcal{X} \subseteq \mathcal{Y}$ *implies* $\emptyset \models \bigsqcap \mathcal{X} \sqsupseteq \bigsqcap \mathcal{Y}$ 5. $\emptyset \models C \sqsubseteq D$ *only if* $\pi_{\mathcal{C}}(C) \supseteq \pi_{\mathcal{C}}(D)$
3. $\mathcal{X} \subseteq \pi_{\mathcal{C}}(\bigsqcap \mathcal{X})$ 6. $\emptyset \models C \sqsubseteq \bigsqcap \pi_{\mathcal{C}}(C)$
4. $\emptyset \models \bigsqcap \mathcal{X} \equiv \bigsqcap \pi_{\mathcal{C}}(\bigsqcap \mathcal{X})$ 7. $\pi_{\mathcal{C}}(C) = \pi_{\mathcal{C}}(\bigsqcap \pi_{\mathcal{C}}(C))$

Proof It suffices to show Statement 1. Then the other statements are obtained as a consequence. We can easily see that the following equivalences hold.

$$\mathcal{X} \subseteq \pi_{\mathcal{C}}(C) \text{ if, and only if, } \forall D \in \mathcal{X} : \emptyset \models C \sqsubseteq D$$

$$\text{if, and only if, } \emptyset \models C \sqsubseteq \bigsqcap \mathcal{X}. \qquad \square$$

In the case of $\mathcal{E}\mathcal{L}^{\perp}_{\text{gfp}}$, Baader and Distel showed that each (unbounded) MMSC of an interpretation $\mathcal{I}$ can be expressed in terms of $\{\perp\} \cup N_C \cup \{\exists r. X^{\mathcal{I}} \mid r \in N_R \text{ and } \emptyset \neq X \subseteq \Delta^{\mathcal{I}}\}$. Similarly, for the role-depth-bounded case, Borchmann, Distel, and Kriegel showed that each RMMSC of $\mathcal{I}$ w.r.t. δ is expressible in terms of $\{\perp\} \cup N_C \cup \{\exists r. X^{\mathcal{I}(\delta-1)} \mid r \in N_R \text{ and } \emptyset \neq X \subseteq \Delta^{\mathcal{I}}\}$. As a straightforward extension to $\mathcal{M}$, we can infer from Theorem 8.3 that each RMMSC is expressible in terms of

$$\mathscr{C}(\mathscr{I},\delta) := \{\bot\} \cup \{A, \neg A \mid A \in N_C\} \cup \left\{ \begin{array}{ll} \forall\, r.\, X^{\mathscr{I}(\delta-1)}, & r \in N_R, \\[2mm] \exists \geq m.\, r.\, X^{\mathscr{I}(\delta-1)}, & 0 < m \leq |\Delta^{\mathscr{I}}|, \\[2mm] \exists \leq n.\, r, & 0 \leq n \leq |\Delta^{\mathscr{I}}|, \\[2mm] \exists\, r.\, \mathsf{Self} & \emptyset \neq X \subseteq \Delta^{\mathscr{I}} \end{array} \right\}$$

$$= \{\bot\} \cup N_C \cup \neg N_C$$

$$\cup\, \forall\, N_R.\, (\mathsf{Mmsc}(\mathscr{I},\delta-1) \setminus \{\bot\})$$

$$\cup\, \exists \geq \{1,\ldots,|\Delta^{\mathscr{I}}|\}.\, N_R.\, (\mathsf{Mmsc}(\mathscr{I},\delta-1) \setminus \{\bot\})$$

$$\cup\, \exists \leq \{0,\ldots,|\Delta^{\mathscr{I}}|\}.\, N_R$$

$$\cup\, \exists\, N_R.\, \mathsf{Self},$$

i.e., the set $\mathscr{C}(\mathscr{I},\delta)$ is $\sqcap$-dense in the set $\mathsf{Mmsc}(\mathscr{I},\delta)$ of all RMMSCs of $\mathscr{I}$ with respect to δ.

Definition 10.2 Let $\mathscr{I}$ be an interpretation, and let $\mathscr{C}$ be a set of $\mathcal{M}$-concept descriptions, both over the same signature Σ. Then, the *induced formal context* of $\mathscr{I}$ and $\mathscr{C}$ is defined as $\mathbb{K}(\mathscr{I},\mathscr{C}) := (\Delta^{\mathscr{I}}, \mathscr{C}, I)$ the incidence of which is defined by $(d, C) \in I$ if, and only if, $d \in C^{\mathscr{I}}$. Furthermore, the *induced formal context* $\mathbb{K}(\mathscr{I},\delta)$ of $\mathscr{I}$ and a role-depth bound $\delta \in \mathbb{N}$ is defined as the induced formal context of $\mathscr{I}$ and $\mathscr{C}(\mathscr{I},\delta)$. The projection $\pi_{\mathscr{C}(\mathscr{I},\delta)}$ with respect to $\mathscr{C}(\mathscr{I},\delta)$ is simply denoted as $\pi_{\mathscr{I},\delta}$.

Lemma 10.3 *Let $\mathbb{K}(\mathscr{I},\mathscr{C})$ be an induced formal context such that $\mathscr{C} \subseteq \mathcal{M}(\Sigma)\!\restriction_\delta$ for a role depth bound $\delta \in \mathbb{N}$. Then, for all subsets $X \subseteq \Delta^{\mathscr{I}}$, all subsets $\mathscr{X} \subseteq \mathscr{C}$, and all $\mathcal{M}$-concept descriptions $C \in \mathcal{M}(\Sigma)$, the following statements hold true.*

1. $\pi_{\mathscr{C}}(X^{\mathscr{I}(\delta)}) = X^{I}$
2. $(\sqcap \mathscr{X})^{\mathscr{I}} = \mathscr{X}^{I}$
3. $C^{\mathscr{I}} \subseteq \pi_{\mathscr{C}}(C)^{I}$
4. $\pi_{\mathscr{C}}((\sqcap \mathscr{X})^{\mathscr{I}\mathscr{I}(\delta)}) = \mathscr{X}^{II}$

Furthermore, if C is expressible in terms of $\mathscr{C}$, then also the following statements are satisfied.

5. $\emptyset \models C \equiv \sqcap \pi_{\mathscr{C}}(C)$
6. $C^{\mathscr{I}} = (\pi_{\mathscr{C}}(C))^{I}$

Eventually, if $\mathscr{X}$ is an intent of $\mathbb{K}(\mathscr{I},\mathscr{C})$, then the following equality is valid, too.

7. $\mathscr{X} = \pi_{\mathscr{C}}(\sqcap \mathscr{X})$

Proof

1. Let $X \subseteq \Delta^{\mathcal{I}}$. Then we have

$$\pi_{\mathscr{C}}(X^{\mathcal{I}(\delta)}) = \{D \in \mathscr{C} \mid \emptyset \models X^{\mathcal{I}(\delta)} \sqsubseteq D\}$$
$$\overset{(*)}{=} \{D \in \mathscr{C} \mid X \subseteq D^{\mathcal{I}}\}$$
$$= \{D \in \mathscr{C} \mid \forall x \in X : (x, D) \in I\}$$
$$= X^{I},$$

where the equality $(*)$ follows from Statement 1 of Lemma 7.3.

2. Let $\mathscr{X} \subseteq \mathscr{C}$. Then it holds that

$$\left(\bigsqcap \mathscr{X}\right)^{\mathcal{I}} = \bigcap \{D^{\mathcal{I}} \mid D \in \mathscr{X}\} = \bigcap \{\{D\}^{I} \mid D \in \mathscr{X}\} = \mathscr{X}^{I}.$$

3. Let $C \in \mathscr{M}(\Sigma)$ be a concept description. Then we have

$$C^{\mathcal{I}} \subseteq \bigcap \{D^{\mathcal{I}} \mid D \in \mathscr{C} \text{ and } \emptyset \models C \sqsubseteq D\}$$
$$= \bigcap \{D^{I} \mid D \in \mathscr{C} \text{ and } \emptyset \models C \sqsubseteq D\}$$
$$= \{D \mid D \in \mathscr{C} \text{ and } \emptyset \models C \sqsubseteq D\}^{I}$$
$$= \pi_{\mathscr{C}}(C)^{I}.$$

4. Let $\mathscr{X} \subseteq \mathscr{C}$ be a set of concept descriptions from $\mathscr{C}$. Then it holds that

$$\pi_{\mathscr{C}}\left(\left(\bigsqcap \mathscr{X}\right)^{\mathcal{I} \mathcal{I}(\delta)}\right) = \{D \in \mathscr{C} \mid \emptyset \models \left(\bigsqcap \mathscr{X}\right)^{\mathcal{I} \mathcal{I}(\delta)} \sqsubseteq D\}$$
$$= \{D \in \mathscr{C} \mid \emptyset \models \left(\bigsqcap \mathscr{X}\right)^{\mathcal{I}} \subseteq D^{\mathcal{I}}\}$$
$$= \{D \in \mathscr{C} \mid \mathscr{X}^{I} \subseteq \{D\}^{I}\}$$
$$= \{D \in \mathscr{C} \mid D \in \mathscr{X}^{II}\}$$
$$= \mathscr{X}^{II}.$$

Now let furthermore C be a concept description that is expressible in terms of $\mathscr{C}$. Then we know that there is a subset $\mathscr{X} \subseteq \mathscr{C}$ such that $\emptyset \models C \equiv \bigsqcap \mathscr{X}$.

5. By an application of Statement 4 of Lemma 10.1 we immediately conclude that

$$\emptyset \models C \equiv \bigsqcap \mathscr{X} \equiv \bigsqcap \pi_{\mathscr{C}}\left(\bigsqcap \mathscr{X}\right) \equiv \bigsqcap \pi_{\mathscr{C}}(C).$$

6. The equality follows from the former Statements 2 and 5—in particular, from $\emptyset \models C \equiv \bigsqcap \pi_{\mathscr{C}}(C)$ we deduce that $C^{\mathscr{I}} = (\bigsqcap \pi_{\mathscr{C}}(C))^{\mathscr{I}} = \pi_{\mathscr{C}}(C)^{I}$.

Finally consider an intent $\mathscr{X}$ of $\mathbb{K}(\mathscr{I}, \mathscr{C})$.

7. We have the following equations which follow from Statement 4 and Statement 7 of Lemma 10.1:

$$\pi_{\mathscr{C}}\left(\bigsqcap \mathscr{X}\right) = \pi_{\mathscr{C}}\left(\bigsqcap \mathscr{X}^{II}\right) = \pi_{\mathscr{C}}\left(\bigsqcap \pi_{\mathscr{C}}\left(\left(\bigsqcap \mathscr{X}\right)^{\mathscr{I} \mathscr{I}(\delta)}\right)\right)$$

$$= \pi_{\mathscr{C}}\left(\left(\bigsqcap \mathscr{X}\right)^{\mathscr{I} \mathscr{I}(\delta)}\right) = \mathscr{X}^{II} = \mathscr{X}. \qquad \square$$

Lemma 10.4 *Let $\mathbb{K}(\mathscr{I}, \mathscr{C})$ be an induced formal context. Then for all subsets $\mathscr{X}, \mathscr{Y} \subseteq \mathscr{C}$, the concept inclusion $\bigsqcap \mathscr{X} \sqsubseteq \bigsqcap \mathscr{Y}$ is valid in $\mathscr{I}$ if, and only if, the implication $\mathscr{X} \to \mathscr{Y}$ is valid in $\mathbb{K}(\mathscr{I}, \mathscr{C})$.*

Proof It is readily verified that the following equivalences hold true.

$$\mathscr{I} \models \bigsqcap \mathscr{X} \sqsubseteq \bigsqcap \mathscr{Y} \text{ if, and only if, } \left(\bigsqcap \mathscr{X}\right)^{\mathscr{I}} \subseteq \left(\bigsqcap \mathscr{Y}\right)^{\mathscr{I}}$$

$$\text{if, and only if, } \mathscr{X}^{I} \subseteq \mathscr{Y}^{I}$$

$$\text{if, and only if, } \mathbb{K}(\mathscr{I}, \mathscr{C}) \models \mathscr{X} \to \mathscr{Y} \qquad \square$$

Definition 10.5 Let $\mathscr{I}$ be an interpretation over the signature Σ, let $\delta \in \mathbb{N}$ be a role depth bound, and assume that C is an $\mathscr{M}$-concept description over Σ. Then the *lower approximation* of C with respect to $\mathscr{I}$ and δ is defined as the concept description

$$\lfloor C \rfloor_{\mathscr{I}, \delta} := \bigsqcap \mathsf{Conj}(C, \{\bot, \top\})$$

$$\sqcap \bigsqcap \mathsf{Conj}(C, N_C)$$

$$\sqcap \bigsqcap \mathsf{Conj}(C, \neg N_C)$$

$$\sqcap \bigsqcap \{\forall r. D^{\mathscr{I} \mathscr{I}(\delta-1)} \mid \forall r. D \in \mathsf{Conj}(C, \forall N_R. \mathscr{M}(\Sigma))\}$$

$$\sqcap \bigsqcap \{\exists \geq n. r. D^{\mathscr{I} \mathscr{I}(\delta-1)} \mid \exists \geq n. r. D \in \mathsf{Conj}(C, \exists \geq \mathbb{N}. N_R. \mathscr{M}(\Sigma))\}$$

$$\sqcap \bigsqcap \mathsf{Conj}(C, \exists \leq \mathbb{N}. N_R)$$

$$\sqcap \bigsqcap \mathsf{Conj}(C, \exists N_R. \mathsf{Self}).$$

Lemma 10.6 *Let $\mathcal{I}$ be an interpretation over the signature Σ, and assume that $\delta \in \mathbb{N}$ is a role depth bound. Then, for all concept descriptions $C, D \in \mathcal{M}(\Sigma)$, all role names $r \in N_R$, and all natural numbers $n \in \mathbb{N}$, the following statements hold true.*

1. $(C \sqcap D)^{\mathcal{I}} = (C^{\mathcal{I}\mathcal{I}(\delta)} \sqcap D)^{\mathcal{I}}$
2. $(\forall r. C)^{\mathcal{I}} = (\forall r. C^{\mathcal{I}\mathcal{I}(\delta)})^{\mathcal{I}}$
3. $(\exists \geq n.r. C)^{\mathcal{I}} = (\exists \geq n.r. C^{\mathcal{I}\mathcal{I}(\delta)})^{\mathcal{I}}$

Proof Beforehand observe that according to Statement 6 of Lemma 7.3, for all $\mathcal{M}$-concept descriptions C over Σ, it holds true that $\emptyset \models C^{\mathcal{I}} \equiv C^{\mathcal{I}\mathcal{I}(\delta)\mathcal{I}}$.

1. It holds true that $(C \sqcap D)^{\mathcal{I}} = C^{\mathcal{I}} \cap D^{\mathcal{I}} = C^{\mathcal{I}\mathcal{I}(\delta)\mathcal{I}} \cap D^{\mathcal{I}} = (C^{\mathcal{I}\mathcal{I}(\delta)} \sqcap D)^{\mathcal{I}}$.

2. It holds true that

$$(\forall r. C)^{\mathcal{I}} = \{d \in \Delta^{\mathcal{I}} \mid \forall e \in \Delta^{\mathcal{I}}: (d, e) \in r^{\mathcal{I}} \text{ implies } e \in C^{\mathcal{I}}\}$$

$$= \{d \in \Delta^{\mathcal{I}} \mid \forall e \in \Delta^{\mathcal{I}}: (d, e) \in r^{\mathcal{I}} \text{ implies } e \in C^{\mathcal{I}\mathcal{I}(\delta)\mathcal{I}}\}$$

$$= (\forall r. C^{\mathcal{I}\mathcal{I}(\delta)})^{\mathcal{I}}.$$

3. It holds true that

$$(\exists \geq n.r. C)^{\mathcal{I}} = \{d \in \Delta^{\mathcal{I}} \mid \exists E \in \binom{\Delta^{\mathcal{I}}}{n} \forall e \in E: (d, e) \in r^{\mathcal{I}} \text{ and } e \in C^{\mathcal{I}}\}$$

$$= \{d \in \Delta^{\mathcal{I}} \mid \exists E \in \binom{\Delta^{\mathcal{I}}}{n} \forall e \in E: (d, e) \in r^{\mathcal{I}} \text{ and } e \in C^{\mathcal{I}\mathcal{I}(\delta)\mathcal{I}}\}$$

$$= (\exists \geq n.r. C^{\mathcal{I}\mathcal{I}(\delta)})^{\mathcal{I}}. \qquad \square$$

Lemma 10.7 *Let $\mathcal{I}$ be an interpretation over Σ. Then for every $\mathcal{M}$-concept description C over Σ the role depth of which does not exceed δ, it holds true that*

$$\emptyset \models \{C^{\mathcal{I}\mathcal{I}(\delta)} \sqsubseteq \lfloor C \rfloor_{\mathcal{I},\delta}, \ \lfloor C \rfloor_{\mathcal{I},\delta} \sqsubseteq C\}.$$

Proof We know that $\emptyset \models D^{\mathcal{I}\mathcal{I}(\delta-1)} \sqsubseteq D$ for all concept descriptions D over Σ with $\mathsf{rd}(D) \leq \delta - 1$, and since value restrictions as well as qualified greater-than restrictions are monotonous in its concept argument, we have that $\emptyset \models \forall r. D^{\mathcal{I}\mathcal{I}(\delta-1)} \sqsubseteq \forall r. D$ and $\emptyset \models \exists \geq n.r. D^{\mathcal{I}\mathcal{I}(\delta-1)} \sqsubseteq \exists \geq n.r. D$ is satisfied for all role names $r \in N_R$ and all natural numbers $n \in \mathbb{N}$. Hence, we conclude that the lower approximation $\lfloor C \rfloor_{\mathcal{I},\delta}$ is subsumed by C with respect to the empty TBox $\emptyset$.

Furthermore, we infer the following equivalences, in particular the equality $(*)$ follows by applying Lemma 10.6.

$$(\lfloor C \rfloor_{\mathcal{I},\delta})^{\mathcal{I}}$$

$$= \left(\bigsqcap \mathsf{Conj}(C, \{\bot, \top\} \cup N_C \cup \neg N_C \cup \exists \leq \mathbb{N}. N_R \cup \exists N_R. \mathsf{Self}) \right.$$

$$\sqcap \prod \{ \forall r. D^{\mathscr{I} \mathscr{I} (\delta-1)} \mid \forall r. D \in \mathsf{Conj}(C, \forall N_R. \mathscr{M}(\Sigma)) \}$$

$$\sqcap \prod \{ \exists \geq n. r. D^{\mathscr{I} \mathscr{I} (\delta-1)} \mid \exists \geq n. r. D \in \mathsf{Conj}(C, \exists \geq \mathbb{N}. N_R. \mathscr{M}(\Sigma)) \} \Big)^{\mathscr{I}}$$

$$= \Big(\prod \mathsf{Conj}(C, \{\bot, \top\} \cup N_C \cup \neg N_C \cup \exists \leq \mathbb{N}. N_R \cup \exists N_R. \mathsf{Self}) \Big)^{\mathscr{I}}$$

$$\sqcap \bigcap \{ (\forall r. D^{\mathscr{I} \mathscr{I} (\delta-1)})^{\mathscr{I}} \mid \forall r. D \in \mathsf{Conj}(C, \forall N_R. \mathscr{M}(\Sigma)) \}$$

$$\sqcap \bigcap \{ (\exists \geq n. r. D^{\mathscr{I} \mathscr{I} (\delta-1)})^{\mathscr{I}} \mid \exists \geq n. r. D \in \mathsf{Conj}(C, \exists \geq \mathbb{N}. N_R. \mathscr{M}(\Sigma)) \}$$

$$\overset{(*)}{=} \Big(\prod \mathsf{Conj}(C, \{\bot, \top\} \cup N_C \cup \neg N_C \cup \exists \leq \mathbb{N}. N_R \cup \exists N_R. \mathsf{Self}) \Big)^{\mathscr{I}}$$

$$\sqcap \bigcap \{ (\forall r. D)^{\mathscr{I}} \mid \forall r. D \in \mathsf{Conj}(C, \forall N_R. \mathscr{M}(\Sigma)) \}$$

$$\sqcap \bigcap \{ (\exists \geq n. r. D)^{\mathscr{I}} \mid \exists \geq n. r. D \in \mathsf{Conj}(C, \exists \geq \mathbb{N}. N_R. \mathscr{M}(\Sigma)) \}$$

$$= C^{\mathscr{I}}$$

Eventually, it follows that $C^{\mathscr{I}} \subseteq (\lfloor C \rfloor_{\mathscr{I}, \delta})^{\mathscr{I}}$ and using Statement 1 of Lemma 7.3 we infer that $\emptyset \models C^{\mathscr{I} \mathscr{I} (\delta)} \sqsubseteq \lfloor C \rfloor_{\mathscr{I}, \delta}$. □

Lemma 10.8 *Let $\mathscr{I}$ be an interpretation and $\delta \in \mathbb{N}$ be a role depth bound. Then every model-based most specific concept description of $\mathscr{I}$ with role depth bound δ is expressible in terms of $\mathscr{C}(\mathscr{I}, \delta)$.*

Proof Let C be a model-based most specific concept description in $\mathscr{I}$ with respect to the role depth δ. Then Statement 3 of Lemma 7.3 yields that $\emptyset \models C \equiv C^{\mathscr{I} \mathscr{I} (\delta)}$. Using the previous Lemma 10.7, we then know that C is equivalent to its lower approximation w.r.t. $\mathscr{I}$. Obviously, C is then expressible in terms of $\mathscr{C}(\mathscr{I}, \delta)$. □

Lemma 10.9 *Let $\mathbb{K}(\mathscr{I}, \delta)$ be an induced formal context. Then, for all subsets $\mathscr{X} \subseteq \mathscr{C}(\mathscr{I}, \delta)$ and all $\mathscr{M}$-concept descriptions C over Σ, the following statements hold true.*

1. $\emptyset \models (\prod \mathscr{X})^{\mathscr{I} \mathscr{I} (\delta)} \equiv \prod \mathscr{X}^{II}$
2. *If $\mathscr{X}$ is an intent of $\mathbb{K}(\mathscr{I}, \delta)$, then $\prod \mathscr{X}$ is a model-based most specific concept description of $\mathscr{I}$ with role-depth bound δ.*
3. *If C is a model-based most specific concept description of $\mathscr{I}$ with role-depth bound δ, then $\pi_{\mathscr{I}, \delta}(C)$ is an intent of $\mathbb{K}(\mathscr{I}, \delta)$.*

Proof

1. We already know that $\mathscr{X}^{II} = \pi_{\mathscr{I}, \delta}((\prod \mathscr{X})^{\mathscr{I} \mathscr{I} (\delta)})$ holds, cf. Statement 4 of Theorem 10.3, and thus also $\emptyset \models \prod \pi_{\mathscr{I}, \delta}((\prod \mathscr{X})^{\mathscr{I} \mathscr{I} (\delta)}) \equiv \prod \mathscr{X}^{II}$. Further-

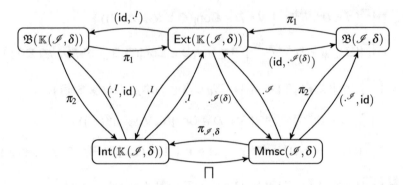

Fig. 4 Overview on the isomorphisms between the extent lattice, intent lattice, and RMMSC lattice of $\mathbb{K}(\mathscr{I}, \delta)$ and $\mathscr{I}$, δ, respectively. Note that $\mathsf{Ext}(\mathbb{K}(\mathscr{I}, \delta)) = \mathsf{Ext}(\mathscr{I}, \delta)$ holds

more, from Lemma 10.8 it follows that $(\bigsqcap \mathscr{X})^{\mathscr{I}\mathscr{I}(\delta)}$ is expressible in terms of $\mathscr{C}(\mathscr{I}, \delta)$, i.e., Statement 5 of Lemma 10.3 implies $\emptyset \models \bigsqcap \pi_{\mathscr{I}, \delta}((\bigsqcap \mathscr{X})^{\mathscr{I}\mathscr{I}(\delta)}) \equiv (\bigsqcap \mathscr{X})^{\mathscr{I}\mathscr{I}(\delta)}$.

2. Let $\mathscr{X} = \mathscr{X}^{II}$ be an intent. Then it follows that $\emptyset \models \bigsqcap \mathscr{X} \equiv \bigsqcap \mathscr{X}^{II}$, and Lemma 10.3 yields $\emptyset \models \bigsqcap \mathscr{X} \equiv (\bigsqcap \mathscr{X})^{\mathscr{I}\mathscr{I}(\delta)}$, i.e., $\bigsqcap \mathscr{X}$ is a RMMSC.

3. Conversely, let C be an RMMSC, i.e., $\emptyset \models C \equiv C^{\mathscr{I}\mathscr{I}(\delta)}$. Then Statement 5 of Lemma 10.3 implies $\emptyset \models C \equiv \bigsqcap \pi_{\mathscr{I}, \delta}(C)$. Furthermore, it follows that $\emptyset \models \bigsqcap \pi_{\mathscr{I}, \delta}(C) \equiv (\bigsqcap \pi_{\mathscr{I}, \delta}(C))^{\mathscr{I}\mathscr{I}(\delta)} \equiv \bigsqcap \pi_{\mathscr{I}, \delta}(C)^{II}$. In particular then $\emptyset \models C \sqsubseteq \bigsqcap \pi_{\mathscr{I}, \delta}(C)^{II}$ holds, and according to Lemma 10.1 this is equivalent to $\pi_{\mathscr{I}, \delta}(C)^{II} \subseteq \pi_{\mathscr{I}, \delta}(C)$. Of course, the inverse set inclusion also holds, i.e., eventually $\pi_{\mathscr{I}, \delta}(C)$ is an intent. □

Corollary 10.10 *The concept lattice of $\mathbb{K}(\mathscr{I}, \delta)$ is isomorphic to the concept lattice of $\mathscr{I}$ and δ. A complete overview on the corresponding isomorphisms is shown in Fig. 4.*

11 Knowledge Bases of Interpretations

In Sect. 4 we introduced the notion of a *concept inclusion*. In particular, a CI $C \sqsubseteq D$ is valid in an interpretation $\mathscr{I}$ if $C^{\mathscr{I}} \subseteq D^{\mathscr{I}}$ is satisfied. We denote the set of all valid CIs of $\mathscr{I}$ by $\mathscr{T}(\mathscr{I})$. In contrast to formal contexts, where there are only finitely many valid implications in case of a finite attribute set, the set $\mathscr{T}(\mathscr{I})$ is infinite, even for finite interpretations over finite signatures. As an example, consider the CI $\top \sqsubseteq \top$, which is valid in all interpretations. Furthermore, if a CI $C \sqsubseteq D$ is valid in $\mathscr{I}$, then so is $\exists r. C \sqsubseteq \exists r. D$. We conclude that $\mathscr{T}(\mathscr{I})$ always contains at least countably infinitely many CIs, provided that there is at least one role name. An important question now is, whether there is a finite *base* of CIs for $\mathscr{I}$, i.e., a (finite) TBox $\mathscr{B}(\mathscr{I})$ such that $\mathscr{B}(\mathscr{I}) \models \mathscr{T}(\mathscr{I})$ as well as $\mathscr{T}(\mathscr{I}) \models \mathscr{B}(\mathscr{I})$. Baader and

Distel found an affirmative answer in [2, 16] for the case of finite interpretations over finite signatures in the description logic $\mathcal{EL}^\perp$, where they take an elegant detour over $\mathcal{EL}^\perp_{\mathsf{gfp}}$, i.e., $\mathcal{EL}^\perp$ interpreted with *greatest fixpoint semantics,* and later Borchmann, Distel, and Kriegel found a positive answer in [12] for finite interpretations over finite signatures in the description logic $\mathcal{EL}^\perp$ restricted by a role depth bound, which is easier to apply and implement, since the *descriptive semantics* are utilized for which plenty of reasoners already exist. Furthermore, it was investigated how the technique of construction of a base of CIs can be iterated for taking into account input interpretations which can be observed on a daily basis, and similarly taking into account existing knowledge in form of a TBox, cf. [29].

Definition 11.1 Let $\mathcal{I}$ be an interpretation over a signature Σ, and assume that $\delta \in \mathbb{N}$ is a role depth bound. Then, a *knowledge base* for $\mathcal{I}$ and δ is a pair $\mathcal{K} := (\mathcal{T}, \mathcal{R})$ consisting of a TBox $\mathcal{T}$ and an RBox $\mathcal{R}$ such that for all concept inclusions α the role depth of the subsumee of which, and of the subsumer of which, respectively, does not exceed δ, and also for all role inclusions α, it holds true that

$$\mathcal{I} \models \alpha \text{ if, and only if, } \mathcal{K} \models \alpha.$$

A knowledge base $\mathcal{K}$ is *non-redundant* if none of the axioms is entailed by the others, i.e., if for each $\alpha \in \mathcal{T} \cup \mathcal{R}$, it holds true that $(\mathcal{T} \setminus \{\alpha\}, \mathcal{R} \setminus \{\alpha\}) \not\models \alpha$. Furthermore, a knowledge base for $\mathcal{I}$ and δ is *minimal* if there is no knowledge base for $\mathcal{I}$ and δ of a smaller cardinality.

By means of the results of the previous sections we are now ready to formulate a knowledge base for an interpretation $\mathcal{I}$, or for a description graph $\mathcal{G}$, respectively. Beforehand, we inspect the interplay of role and concept inclusions, and we list some trivial concept inclusions that are valid in all interpretations.

Lemma 11.2 *Let $m, n \in \mathbb{N}_+$ be non-negative integers with $n < m$, $r \in N_R$ be a role name, and C, D be $\mathcal{M}$-concept descriptions. Then, the following concept inclusions hold in every interpretation $\mathcal{I}$.*

$$A \sqcap \neg A \sqsubseteq \bot$$

$$\exists r.\, \textit{Self} \sqcap \forall r.\, C \sqsubseteq C$$

$$\exists r.\, \textit{Self} \sqcap C \sqsubseteq \exists r.\, C$$

$$\exists r.\, \textit{Self} \sqcap C \sqcap \exists \leq 1.\, r \sqsubseteq \forall r.\, C$$

$$\exists \geq n.\, r.\, C \sqcap \forall r.\, D \sqsubseteq \exists \geq n.\, r.\, (C \sqcap D)$$

$$\exists \leq n.\, r \sqsubseteq \exists \leq m.\, r$$

$$\exists \geq m.\, r.\, C \sqsubseteq \exists \geq n.\, r.\, C$$

$$\exists \geq |\Delta^{\mathcal{I}}|.\, r.\, C \sqsubseteq C \sqcap \forall r.\, C \sqcap \exists r.\, \textit{Self}$$

$$\top \sqsubseteq \exists \leq |\Delta^{\mathcal{I}}|.\, r$$

Proof Most of the concept inclusions are obviously valid. We are only going to explain the validity of the penultimate concept inclusion. If a domain element has at least $|\Delta^{\mathcal{I}}|$ r-successors in C, then especially it must be an r-successor of itself, hence be in C and in $\exists r.\,\mathsf{Self}$. Furthermore, there cannot be any further r-successors, and so all r-successors must be in C. □

Please note that there are no direct subsumptions between existential restrictions $\exists r.\,C$ and value restrictions $\forall r.\,C$, i.e., both $\exists r.\,C \sqsubseteq \forall r.\,C$ and $\forall r.\,C \sqsubseteq \exists r.\,C$ do not hold. There is also a crossover between both which is denoted by $\forall\exists$, and has the semantics $(\forall\exists r.\,C)^{\mathcal{I}} := (\exists r.\,C)^{\mathcal{I}} \cap (\forall r.\,C)^{\mathcal{I}}$, i.e., a domain element is in the extension of $\forall\exists r.\,C$ if, and only if, there is an r-successor in C, and all r-successors are in C. Furthermore, there is also a *reversed* value restriction $\forall C.\,r$ with the semantics $(\forall C.\,r)^{\mathcal{I}} := \{ d \in \Delta^{\mathcal{I}} \mid \forall e \in \Delta^{\mathcal{I}}: e \in C^{\mathcal{I}}$ implies $(d,e) \in r^{\mathcal{I}} \}$. However, we do not use either of them for our mining technique.

The next two lemmas show us which concept inclusions can be inferred from known role inclusions.

Lemma 11.3 *Let $\mathcal{I}$ be a model of the role inclusion $r \sqsubseteq s$, as well as of the concept inclusion $C \sqsubseteq D$, and furthermore let $m \leq n$ be natural numbers. Then $\mathcal{I}$ is also a model of the following concept inclusions.*

$$\exists \geq n.\,r.\,C \sqsubseteq \exists \geq m.\,s.\,D$$

$$\exists r.\,\mathsf{Self} \sqsubseteq \exists s.\,\mathsf{Self}$$

$$\forall s.\,C \sqsubseteq \forall r.\,D$$

$$\exists \leq m.\,s \sqsubseteq \exists \leq n.\,r$$

Proof Assume that $m \leq n$, and let $\mathcal{I}$ be an interpretation such that $r^{\mathcal{I}} \subseteq s^{\mathcal{I}}$ and $C^{\mathcal{I}} \subseteq D^{\mathcal{I}}$.

($\geq$) Then we have that

$$(\exists \geq n.\,r.\,C)^{\mathcal{I}} = \{ d \in \Delta^{\mathcal{I}} \mid \exists E \in \binom{\Delta^{\mathcal{I}}}{n}: \{d\} \times E \subseteq r^{\mathcal{I}} \text{ and } E \subseteq C^{\mathcal{I}} \}$$

$$\subseteq \{ d \in \Delta^{\mathcal{I}} \mid \exists E \in \binom{\Delta^{\mathcal{I}}}{m}: \{d\} \times E \subseteq s^{\mathcal{I}} \text{ and } E \subseteq D^{\mathcal{I}} \}$$

$$= (\exists \geq m.\,s.\,D)^{\mathcal{I}}.$$

($\exists$) For the existential self restrictions we can infer the following.

$$(\exists r.\,\mathsf{Self})^{\mathcal{I}} = \{ d \in \Delta^{\mathcal{I}} \mid (d,d) \in r^{\mathcal{I}} \}$$

$$\subseteq \{ d \in \Delta^{\mathcal{I}} \mid (d,d) \in s^{\mathcal{I}} \}$$

$$= (\exists s.\,\mathsf{Self})^{\mathcal{I}}$$

(∀)   Furthermore, consider a concept inclusion $\forall s.\, C \sqsubseteq \forall r.\, C$. We can infer the following.

$$(\forall s.\, C)^{\mathscr{I}} = \{ d \in \Delta^{\mathscr{I}} \mid \forall e \in \Delta^{\mathscr{I}} : (d, e) \in s^{\mathscr{I}} \text{ implies } e \in C^{\mathscr{I}} \}$$

$$\subseteq \{ d \in \Delta^{\mathscr{I}} \mid \forall e \in \Delta^{\mathscr{I}} : (d, e) \in r^{\mathscr{I}} \text{ implies } e \in D^{\mathscr{I}} \}$$

$$= (\forall r.\, C)^{\mathscr{I}}$$

(≤)   Finally, it holds true that

$$(\exists \leq m.\, s)^{\mathscr{I}} = \{ d \in \Delta^{\mathscr{I}} \mid \forall E \in \binom{\Delta^{\mathscr{I}}}{m+1} : \{d\} \times E \nsubseteq s^{\mathscr{I}} \}$$

$$\subseteq \{ d \in \Delta^{\mathscr{I}} \mid \forall E \in \binom{\Delta^{\mathscr{I}}}{n+1} : \{d\} \times E \nsubseteq r^{\mathscr{I}} \}$$

$$= (\exists \leq n.\, r)^{\mathscr{I}}. \qquad\qquad \square$$

First, we want to extract a minimal RBox $\mathscr{R}(\mathscr{I})$ from the interpretation that entails all role inclusions valid in $\mathscr{I}$. We therefore define an equivalence relation $\equiv_{\mathscr{I}}$ on the role names as follows: $r \equiv_{\mathscr{I}} s$ if, and only if, $r^{\mathscr{I}} = s^{\mathscr{I}}$. Then let $N_R^{\mathscr{I}}$ be a set of representatives of this equivalence relation, i.e., $|N_R^{\mathscr{I}} \cap [r]_{\equiv_{\mathscr{I}}}| = 1$ for all role names $r \in N_R$. If $[r]_{\equiv_{\mathscr{I}}} = \{r_1, \ldots, r_\ell\}$ is an enumeration of the equivalence class of r, then add the following role equivalence axioms to $\mathscr{R}(\mathscr{I})$.

$$\mathscr{R}(\mathscr{I}, r) := \{ r_1 \sqsubseteq r_2, r_2 \sqsubseteq r_3, \ldots, r_{\ell-1} \sqsubseteq r_\ell, r_\ell \sqsubseteq r_1 \}$$

Furthermore, define an order relation $\sqsubseteq_{\mathscr{I}}$ on the representatives $N_R^{\mathscr{I}}$ by $r \sqsubseteq_{\mathscr{I}} s$ if, and only if, $r^{\mathscr{I}} \subseteq s^{\mathscr{I}}$. Let $\prec_{\mathscr{I}}$ be the neighborhood relation of $\sqsubseteq_{\mathscr{I}}$, then add the role inclusion axioms $r \sqsubseteq s$ for each pair $r \prec_{\mathscr{I}} s$ to the RBox $\mathscr{R}(\mathscr{I})$. Obviously, the constructed RBox is minimal w.r.t. the property to entail all valid role inclusion axioms holding in the interpretation $\mathscr{I}$. Eventually, the RBox is defined as follows.

$$\mathscr{R}(\mathscr{I}) := \{ r \sqsubseteq s \mid r, s \in N_R^{\mathscr{I}} \text{ and } r \prec_{\mathscr{I}} s \} \cup \bigcup \{ \mathscr{R}(\mathscr{I}, r) \mid r \in N_R^{\mathscr{I}} \}$$

Proposition 11.4 *Let $\mathscr{I}$ be an interpretation. Then the RBox $\mathscr{R}(\mathscr{I})$ as defined above is a base for the role inclusions which are valid in $\mathscr{I}$, i.e., for each role inclusion $r \sqsubseteq s$, the following equivalence holds true.*

$$\mathscr{I} \models r \sqsubseteq s \text{ if, and only if, } \mathscr{R}(\mathscr{I}) \models r \sqsubseteq s$$

In particular, $\mathscr{R}(\mathscr{I})$ is non-redundant, i.e., for every role inclusion $r \sqsubseteq s \in \mathscr{R}(\mathscr{I})$, it holds true that $\mathscr{R}(\mathscr{I}) \setminus \{r \sqsubseteq s\} \not\models r \sqsubseteq s$.

Proof The statements are immediate consequences of the construction of $\mathscr{R}(\mathscr{I})$ preceding the proposition.   □

Lemma 11.5 *Let $\mathscr{I}$ be an interpretation over a signature Σ, let C and D be $\mathscr{M}$-concept descriptions over Σ, and further assume that $\delta \in \mathbb{N}$ is a role depth bound. If the CI $C \sqsubseteq D$ is valid in $\mathscr{I}$, and both C and D have a role depth not exceeding δ, then the CI $C \sqsubseteq C^{\mathscr{I}\mathscr{I}(\delta)}$ is valid in $\mathscr{I}$ too, and furthermore, $C \sqsubseteq D$ follows from $C \sqsubseteq C^{\mathscr{I}\mathscr{I}(\delta)}$.*

Proof For the concept description C it follows by an application of Statement 6 of Lemma 7.3 that $C^{\mathscr{I}} = C^{\mathscr{I}\mathscr{I}(\delta)\mathscr{I}}$, i.e., the CI $C \sqsubseteq C^{\mathscr{I}\mathscr{I}(\delta)}$ is always valid in $\mathscr{I}$.

Now consider a model $\mathscr{J}$ of the CI $C \sqsubseteq C^{\mathscr{I}\mathscr{I}(\delta)}$. Since $\mathscr{I} \models C \sqsubseteq D$, it follows that $C^{\mathscr{I}} \subseteq D^{\mathscr{I}}$, and by Statement 1 of Lemma 7.3 we conclude that $\emptyset \models C^{\mathscr{I}\mathscr{I}(\delta)} \sqsubseteq D$. In particular, then the last CI is also valid in $\mathscr{J}$, and hence $\mathscr{J} \models C \sqsubseteq D$. Since $\mathscr{J}$ was an arbitrary model, we conclude that $\{C \sqsubseteq C^{\mathscr{I}\mathscr{I}(\delta)}\} \models C \sqsubseteq D$.   □

Proposition 11.6 *Let $\mathscr{I}$ be a finite interpretation, and let $\delta \in \mathbb{N}$ be a role depth bound. Then, the following TBox is sound and complete for the CIs which satisfy the role depth bound δ and are valid in $\mathscr{I}$.*

$$\{\textstyle\bigsqcap \mathscr{X} \sqsubseteq \bigsqcap \mathscr{X}^{II} \mid \mathscr{X} \subseteq \mathscr{C}(\mathscr{I},\delta)\}$$

$$\cup \{\exists \geq (|\Delta^{\mathscr{I}}| + 1).r.\top \sqsubseteq \bot, \top \sqsubseteq \exists \leq |\Delta^{\mathscr{I}}|.r \mid r \in N_R\}$$

Proof For the sake of improving the readability, denote the above given TBox as $\mathscr{T}$. Since for all $\mathscr{X} \subseteq \mathscr{C}(\mathscr{I},\delta)$, the implication $\mathscr{X} \to \mathscr{X}^{II}$ trivially holds in the induced formal context $\mathbb{K}(\mathscr{I},\delta)$, it immediately follows by an application of Lemma 10.4 that the CI $\bigsqcap \mathscr{X} \sqsubseteq \bigsqcap \mathscr{X}^{II}$ is valid in $\mathscr{I}$. Consequently, we have just proven the soundness of $\mathscr{T}$.

Consider a CI $C \sqsubseteq D$ which is valid in $\mathscr{I}$, and where both C and D possess a role depth of at most δ. Then Lemma 11.5 yields that the CI $C \sqsubseteq C^{\mathscr{I}\mathscr{I}(\delta)}$ is also valid in $\mathscr{I}$, and furthermore the entailment $\{C \sqsubseteq C^{\mathscr{I}\mathscr{I}(\delta)}\} \models C \sqsubseteq D$ holds true. Hence, it suffices to show that our TBox $\mathscr{T}$ entails all CIs of the form $C \sqsubseteq C^{\mathscr{I}\mathscr{I}(\delta)}$. For this purpose, consider an arbitrary model $\mathscr{J}$ of $\mathscr{T}$ as well as an arbitrary concept description $C \in \mathscr{M}(\Sigma)\!\restriction_{\delta}$—we are now going to prove that the CI $C \sqsubseteq C^{\mathscr{I}\mathscr{I}(\delta)}$ is valid in $\mathscr{J}$, too. Beforehand, note that for the right-hand sides of the CIs it holds true that $\emptyset \models \bigsqcap \mathscr{X}^{II} \equiv (\bigsqcap \mathscr{X})^{\mathscr{I}\mathscr{I}(\delta)}$, cf. Statement 1 of Lemma 10.9. Furthermore, we also know that each CI $C \sqsubseteq C^{\mathscr{I}\mathscr{I}(\delta)}$ where C is expressible in terms of $\mathscr{C}(\mathscr{I},\delta)$ is valid in $\mathscr{J}$. We prove this as follows: if C is expressible in terms of $\mathscr{C}(\mathscr{I},\delta)$, then there is a subset $\mathscr{X} \subseteq \mathscr{C}(\mathscr{I},\delta)$ such that $\emptyset \models C \equiv \bigsqcap \mathscr{X}$. Since $\mathscr{J} \models \bigsqcap \mathscr{X} \sqsubseteq (\bigsqcap \mathscr{X})^{\mathscr{I}\mathscr{I}(\delta)}$, we can immediately conclude that $\mathscr{J} \models C \sqsubseteq C^{\mathscr{I}\mathscr{I}(\delta)}$.

We proceed with a proof by induction on the structure of C.

Let $C = \bot$. Since $\bot \in \mathscr{C}(\mathscr{I},\delta)$, we may immediately conclude that $\mathscr{J} \models \bot \sqsubseteq \bot^{\mathscr{I}\mathscr{I}(\delta)}$.

Assume that $C = \top$. From $\top = \bigsqcap \emptyset$ it follows that $\mathscr{J} \models \top \sqsubseteq \top^{\mathscr{I}\mathscr{I}(\delta)}$.

For a concept name $C = A \in N_C$, we have that $A \in \mathscr{C}(\mathscr{I}, \delta)$, and hence $\mathscr{J} \models A \sqsubseteq A^{\mathscr{I}\mathscr{I}(\delta)}$.

For a primitive negation $C = \neg A$, it follows that $\neg A \in \mathscr{C}(\mathscr{I}, \delta)$, and so we conclude that $\mathscr{J} \models \neg A \sqsubseteq (\neg A)^{\mathscr{I}\mathscr{I}(\delta)}$.

Consider a conjunction $C = D \sqcap E$. By induction hypothesis it holds true that $\mathscr{J} \models D \sqsubseteq D^{\mathscr{I}\mathscr{I}(\delta)}$ as well as $\mathscr{J} \models E \sqsubseteq E^{\mathscr{I}\mathscr{I}(\delta)}$. Consequently,

$$\mathscr{J} \models D \sqcap E \sqsubseteq D^{\mathscr{I}\mathscr{I}(\delta)} \sqcap E^{\mathscr{I}\mathscr{I}(\delta)}$$
$$\sqsubseteq (D^{\mathscr{I}\mathscr{I}(\delta)} \sqcap E^{\mathscr{I}\mathscr{I}(\delta)})^{\mathscr{I}\mathscr{I}(\delta)}$$
$$\sqsubseteq (D \sqcap E)^{\mathscr{I}\mathscr{I}(\delta)}.$$

The second subsumption follows from the fact that the concept description $D^{\mathscr{I}\mathscr{I}(\delta)} \sqcap E^{\mathscr{I}\mathscr{I}(\delta)}$ is expressible in terms of $\mathscr{C}(\mathscr{I}, \delta)$, and the last subsumption is a consequence of Statement 5 of Lemma 7.3.

Assume that $C = \forall r. D$ is a value restriction. Then the following subsumptions hold true in $\mathscr{J}$.

$$\mathscr{J} \models \forall r. D \sqsubseteq \forall r. D^{\mathscr{I}\mathscr{I}(\delta)}$$
$$\sqsubseteq \forall r. D^{\mathscr{I}\mathscr{I}(\delta-1)}$$
$$\sqsubseteq (\forall r. D^{\mathscr{I}\mathscr{I}(\delta-1)})^{\mathscr{I}\mathscr{I}(\delta)}$$
$$\sqsubseteq (\forall r. D)^{\mathscr{I}\mathscr{I}(\delta)}$$

The first subsumption is a consequence of the induction hypothesis and the fact that value restrictions are monotonous. For the second subsumption, observe that $D^{\mathscr{I}\mathscr{I}(\delta-1)}$ certainly satisfies that $\mathsf{rd}(D^{\mathscr{I}\mathscr{I}(\delta-1)}) \le \delta$ as well as $D^{\mathscr{I}} \subseteq D^{\mathscr{I}\mathscr{I}(\delta-1)\mathscr{I}}$, and so an application of Statement 3 of Definition 7.1 yields that $\emptyset \models D^{\mathscr{I}\mathscr{I}(\delta)} \sqsubseteq D^{\mathscr{I}\mathscr{I}(\delta-1)}$. Since $\forall r. D^{\mathscr{I}\mathscr{I}(\delta-1)}$ is contained in $\mathscr{C}(\mathscr{I}, \delta)$, it must in particular be expressible in terms of $\mathscr{C}(\mathscr{I}, \delta)$, and this justifies the validity of the third subsumption. Again, the last subsumption follows from Statement 5 of Lemma 7.3.

Now let $C = \exists \ge n. r. D$ be a qualified greater-than restriction, and first assume that $n \le |\Delta^{\mathscr{I}}|$. Then, we may argue similarly as for the value restrictions that the following subsumptions hold true in $\mathscr{J}$.

$$\mathscr{J} \models \exists \ge n. r. D \sqsubseteq \exists \ge n. r. D^{\mathscr{I}\mathscr{I}(\delta)}$$
$$\sqsubseteq \exists \ge n. r. D^{\mathscr{I}\mathscr{I}(\delta-1)}$$
$$\sqsubseteq (\exists \ge n. r. D^{\mathscr{I}\mathscr{I}(\delta-1)})^{\mathscr{I}\mathscr{I}(\delta)}$$
$$\sqsubseteq (\exists \ge n. r. D)^{\mathscr{I}\mathscr{I}(\delta)}$$

For the remaining case where $n > |\Delta^{\mathscr{I}}|$, we argue as follows:

$$\mathscr{I} \models \exists \geq n.r.D \sqsubseteq \exists \geq n.r.\top$$
$$\sqsubseteq \exists \geq |\Delta^{\mathscr{I}}| + 1.r.\top$$
$$\sqsubseteq \bot,$$

and hence the concept descriptions $\bot$ and $\exists \geq n.r.D$ are equivalent in $\mathscr{I}$. Since we have already proven above that $\bot \sqsubseteq \bot^{\mathscr{I}\mathscr{I}(\delta)}$ is valid in $\mathscr{I}$, also the CI $\exists \geq n.r.D \sqsubseteq (\exists \geq n.r.D)^{\mathscr{I}\mathscr{I}(\delta)}$ is valid in $\mathscr{I}$.

Assume that $C = \exists \leq n.r$ is an unqualified less-than restriction, and let $n \leq |\Delta^{\mathscr{I}}|$. Of course, then $\mathscr{I} \models \exists \leq n.r \sqsubseteq (\exists \leq n.r)^{\mathscr{I}\mathscr{I}(\delta)}$ certainly holds true, since $\exists \leq n.r \in \mathscr{C}(\mathscr{I}, \delta)$. In case $n > |\Delta^{\mathscr{I}}|$, then $\exists \leq n.r$ and $\top$ are equivalent in $\mathscr{I}$, and the validity of $\mathscr{I} \models \exists \leq n.r \sqsubseteq (\exists \leq n.r)^{\mathscr{I}\mathscr{I}(\delta)}$ follows from $\mathscr{I} \models \top \sqsubseteq \top^{\mathscr{I}\mathscr{I}(\delta)}$, which we have shown above.

Eventually, consider an existential self restriction $\exists r.\mathsf{Self}$. Obviously, $\exists r.\mathsf{Self}$ is contained in $\mathscr{C}(\mathscr{I}, \delta)$, and so the CI $\exists r.\mathsf{Self} \sqsubseteq (\exists r.\mathsf{Self})^{\mathscr{I}\mathscr{I}(\delta)}$ is valid in $\mathscr{I}$.

□

As final step we use the trivial concept inclusions and concept inclusions that are entailed by valid role inclusions to define some background knowledge for the computation of the canonical implication base of the induced concept context which is trivial in terms of Description Logics, but not for Formal Concept Analysis, due to their different semantics.

Theorem 11.7 *Let $\mathscr{I}$ be an interpretation over the signature Σ, and $\delta \in \mathbb{N}$ a role-depth bound. Furthermore, assume that $\mathscr{L}$ is an implication base of the induced formal context $\mathbb{K}(\mathscr{I}, \delta)$ with respect to the background knowledge*

$$\mathscr{S}(\mathscr{I}, \delta) := \left\{ \{C_1, \ldots, C_\ell\} \to \{D\} \;\middle|\; \begin{array}{l} C_1, \ldots, C_\ell, D \in \mathscr{C}(\mathscr{I}, \delta) \\ \text{and } \mathscr{R}(\mathscr{I}) \models C_1 \sqcap \ldots \sqcap C_\ell \sqsubseteq D \end{array} \right\}.$$

Then $((\bigsqcap \mathscr{L}) \cup \mathscr{N}(\mathscr{I}), \mathscr{R}(\mathscr{I}))$ where

$$\mathscr{N}(\mathscr{I}) := \{\exists \geq (|\Delta^{\mathscr{I}}| + 1).r.\top \sqsubseteq \bot, \top \sqsubseteq \exists \leq |\Delta^{\mathscr{I}}|.r \mid r \in N_R\}$$

is a knowledge base for $\mathscr{I}$ and δ. In particular, the canonical knowledge base for $\mathscr{I}$ and δ is defined as

$$\mathscr{K}(\mathscr{I}, \delta) := (\mathscr{T}(\mathscr{I}, \delta) \cup \mathscr{N}(\mathscr{I}), \mathscr{R}(\mathscr{I}))$$

where $\mathscr{T}(\mathscr{I}, \delta) := \{\bigsqcap \mathscr{P} \sqsubseteq \bigsqcap \mathscr{P}'' \mid \mathscr{P} \in \mathsf{PsInt}(\mathbb{K}(\mathscr{I}, \delta), \mathscr{S}(\mathscr{I}, \delta))\}$.

Proof It is obvious that

$$\mathscr{K}(\mathscr{I}, \delta) = ((\bigsqcap \mathsf{Can}(\mathbb{K}(\mathscr{I}, \delta), \mathscr{S}(\mathscr{I}, \delta))) \cup \mathscr{N}(\mathscr{I}), \mathscr{R}(\mathscr{I})),$$

and hence it suffices to prove that for each implication base $\mathcal{L}$ of $\mathbb{K}(\mathcal{I}, \delta)$ with respect to the background knowledge $\mathcal{S}(\mathcal{I}, \delta)$, the pair $\mathcal{K} := ((\bigsqcap \mathcal{L}) \cup \mathcal{N}(\mathcal{I}), \mathcal{R}(\mathcal{I}))$ is a knowledge base for $\mathcal{I}$.

It is obvious that $\mathcal{I} \models \mathcal{K}$, i.e., $\mathcal{K}$ is sound. We proceed with proving completeness. Completeness for role inclusions follows immediately from Proposition 11.4. In Proposition 11.6 we have proven that the TBox

$$\{\bigsqcap \mathcal{X} \sqsubseteq \bigsqcap \mathcal{X}^{II} \mid \mathcal{X} \subseteq \mathcal{C}(\mathcal{I}, \delta)\} \cup \mathcal{N}(\mathcal{I})$$

is complete for the concept inclusions which are valid in $\mathcal{I}$ and satisfy the role depth bound δ, and thus it suffices to show that for each subset $\mathcal{X} \subseteq \mathcal{C}(\mathcal{I}, \delta)$,

$$\mathcal{K} \models \bigsqcap \mathcal{X} \sqsubseteq \bigsqcap \mathcal{X}^{II}.$$

Consider a model $\mathcal{J}$ of $\mathcal{K}$. We divide the remaining part of this proof in three steps:

1. First, we show that all implications in $\mathcal{L}$ are also valid in the induced formal context $\mathbb{K}(\mathcal{J}, \mathcal{C}(\mathcal{I}, \delta))$ the incidence relation of which we denote as J.
2. Then, we prove that the background knowledge $\mathcal{S}(\mathcal{I}, \delta)$ is valid in the induced formal context $\mathbb{K}(\mathcal{J}, \mathcal{C}(\mathcal{I}, \delta))$, too.
3. Finally, we show that $\mathcal{J}$ is a model of the CI $\bigsqcap \mathcal{X} \sqsubseteq \bigsqcap \mathcal{X}^{II}$.

From the last step, we then immediately conclude that $\mathcal{J}$ is also a model of the TBox from Proposition 11.6. Since $\mathcal{J}$ was chosen arbitrarily, then $\mathcal{K}$ must be complete.

W.l.o.g. we may assume that $\mathcal{L}$ only contains implications of the form $\mathcal{X} \rightarrow \mathcal{X}^{II}$. Hence, let $\mathcal{X} \rightarrow \mathcal{X}^{II} \in \mathcal{L}$, then it follows that

$$\mathcal{X}^J = (\bigsqcap \mathcal{X})^{\mathcal{J}} \subseteq (\bigsqcap \mathcal{X}^{II})^{\mathcal{J}} = \mathcal{X}^{IIJ},$$

i.e., the implication $\mathcal{X} \rightarrow \mathcal{X}^{II}$ is valid in $\mathbb{K}(\mathcal{J}, \mathcal{C}(\mathcal{I}, \delta))$.

Now consider an implication $\{C_1, \dots, C_\ell\} \rightarrow \{D\}$ in $\mathcal{S}(\mathcal{I}, \delta)$, i.e., it holds true that $C_1, \dots, C_\ell, D \in \mathcal{C}(\mathcal{I}, \delta)$ and $\mathcal{R}(\mathcal{I}) \models C_1 \sqcap \dots \sqcap C_\ell \sqsubseteq D$. Since $\mathcal{J}$ is a model of $\mathcal{R}(\mathcal{I})$, the aforementioned CI is valid in $\mathcal{J}$. Lemma 10.4 then justifies that the considered implication must be valid in the induced formal context $\mathbb{K}(\mathcal{J}, \mathcal{C}(\mathcal{I}, \delta))$.

As the last step, we consider an arbitrary CI $\bigsqcap \mathcal{X} \sqsubseteq \bigsqcap \mathcal{X}^{II}$ where $\mathcal{X} \subseteq \mathcal{C}(\mathcal{I}, \delta)$, and prove that it is valid in $\mathcal{J}$. Since the implication set $\mathcal{L} \cup \mathcal{S}(\mathcal{I}, \delta)$ is sound and complete for $\mathbb{K}(\mathcal{I}, \delta)$, and $\mathcal{X} \rightarrow \mathcal{X}^{II}$ is trivially valid in $\mathbb{K}(\mathcal{I}, \delta)$, it holds true that $\mathcal{X} \rightarrow \mathcal{X}^{II}$ is entailed by $\mathcal{L} \cup \mathcal{S}(\mathcal{I}, \delta)$. Consequently, since $\mathbb{K}(\mathcal{J}, \mathcal{C}(\mathcal{I}, \delta))$ is a model of both $\mathcal{L}$ and $\mathcal{S}(\mathcal{I}, \delta)$, it follows that $\mathcal{X} \rightarrow \mathcal{X}^{II}$ is valid in $\mathbb{K}(\mathcal{J}, \mathcal{C}(\mathcal{I}, \delta))$, too. By Lemma 10.4 we conclude that the CI $\bigsqcap \mathcal{X} \sqsubseteq \bigsqcap \mathcal{X}^{II}$ is valid in $\mathcal{J}$. $\qquad \square$

Constructor	$\mathcal{EL}$	$\mathcal{FL}_0$	$\mathcal{FLE}$	$\mathcal{ALE}$	$\mathcal{MH}$
$\bot$				×	×
$\top$	×	×	×	×	×
$\neg A$				×	×
$C \sqcap D$	×	×	×	×	×
$\exists r.C$	×		×	×	×
$\forall r.C$		×	×	×	×
$\exists \geq n.r.C$					×
$\exists \leq n.r$					×
$\exists r.\,\mathsf{Self}$					×
$C \sqsubseteq D$	×	×	×	×	×
$C \equiv D$	×	×	×	×	×
$r \sqsubseteq s$					×

Fig. 5 Overview on various Description Logics below $\mathcal{MH}$

12 Other Description Logics

If only a lower expressivity of the underlying description logic is necessary, then one could also use $\mathcal{EL}$, $\mathcal{FL}_0$, $\mathcal{FLE}$, $\mathcal{ALE}$, or extensions thereof with role hierarchies $\mathcal{H}$. All of the previous results are then still valid, if the expressivity is not higher than that of $\mathcal{MH}$. Figure 5 gives an overview on description logics that have a lower expressivity than $\mathcal{MH}$, and can thus also be used for knowledge acquisition.

As a future step, it would be interesting to investigate methods that also take into account complex role inclusions, e.g., consider the description logic $\mathcal{MR}$. A *complex role inclusion* is an expression $r_1 \circ \ldots \circ r_n \sqsubseteq s$ where $r_1, \ldots, r_n, s \in N_R$ are role names. Its semantics is defined by

$$\mathcal{I} \models r_1 \circ \ldots \circ r_n \sqsubseteq s \text{ if, and only if, } r_1^{\mathcal{I}} \circ \ldots \circ r_n^{\mathcal{I}} \subseteq s^{\mathcal{I}},$$

where $\circ$ denotes composition of binary relations, i.e.,

$$r_1^{\mathcal{I}} \circ \ldots \circ r_n^{\mathcal{I}} = \left\{ (d_0, d_n) \in \Delta^{\mathcal{I}} \times \Delta^{\mathcal{I}} \;\middle|\; \begin{array}{l} \exists d_1, \ldots, d_{n-1} \in \Delta^{\mathcal{I}}: \\ (d_0, d_1) \in r_1^{\mathcal{I}}, \ldots, (d_{n-1}, d_n) \in r_n^{\mathcal{I}} \end{array} \right\}.$$

13 Conclusion

We have provided an extension of the results of Baader and Distel [2, 3, 16] for the deduction of knowledge bases from interpretations in the more expressive description logic $\mathcal{MH}$ w.r.t. descriptive semantics and role-depth bounds, and furthermore explained how this technique can be applied to social graphs. Since role-depth-bounded model-based most specific concept descriptions always exist, this technique can always be applied. Furthermore, the construction of knowledge bases has been reduced to the computation of implication bases of formal contexts, which is a well-understood problem that has several available algorithms—for example the standard *NextClosure* algorithm by Ganter [21, 23], or the parallel algorithm *NextClosures* that was introduced in [28, 30–32] and implemented in [27]. The presented methods in this document are also prototypically implemented in *Concept Explorer FX* [27].

References

1. Baader, F.: Least common subsumers and most specific concepts in a description logic with existential restrictions and terminological cycles. In: Gottlob, G., Walsh, T. (eds.) IJCAI-03, Proceedings of the Eighteenth International Joint Conference on Artificial Intelligence, pp. 319–324, Acapulco, 9–15 August 2003. Morgan Kaufmann, San Francisco (2003). http://dblp.uni-trier.de/rec/bib/conf/ijcai/Baader03

2. Baader, F., Distel, F.: A finite basis for the set of $\mathcal{EL}$-implications holding in a finite model. In: Medina, R., Obiedkov, S.A. (eds.) Formal Concept Analysis, Proceedings of the 6th International Conference, ICFCA 2008, Montreal, 25–28 February 2008. Lecture Notes in Computer Science, vol. 4933, pp. 46–61. Springer, Heidelberg (2008). doi:10.1007/978-3-540-78137-0_4. http://dblp.uni-trier.de/rec/bib/conf/icfca/BaaderD08

3. Baader, F., Distel, F.: Exploring finite models in the description logic $\mathcal{EL}_{gfp}$. In: Ferré, S., Rudolph, S. (eds.) Formal Concept Analysis, Proceedings of the 7th International Conference, ICFCA 2009, Darmstadt, 21–24 May 2009. Lecture Notes in Computer Science, vol. 5548, pp. 146–161. Springer, Heidelberg (2009). doi:10.1007/978-3-642-01815-2_12. http://dblp.uni-trier.de/rec/bib/conf/icfca/BaaderD09

4. Baader, F., Küsters, R., Molitor, R.: Computing least common subsumers in description logics with existential restrictions. In: Dean, T. (ed.) Proceedings of the Sixteenth International Joint Conference on Artificial Intelligence, IJCAI 99, pp. 96–103, Stockholm, 31 July–6 August 1999. 2 Volumes, 1450 pages. Morgan Kaufmann, San Francisco (1999). http://dblp.uni-trier.de/rec/bib/conf/ijcai/BaaderKM99

5. Baader, F., Horrocks, I., Sattler, U.: Description logics. In: van Harmelen, F., Lifschitz, V., Porter, B.W. (eds.) Handbook of Knowledge Representation. Foundations of Artificial Intelligence, vol. 3, pp. 135–179. Elsevier, Amsterdam (2008). doi:10.1016/S1574-6526(07)03003-9. http://dblp.uni-trier.de/rec/bib/reference/fai/3

6. Babin, M.A., Kuznetsov, S.O.: Recognizing pseudo-intents is coNP-complete. In: Kryszkiewicz, M., Obiedkov, S.A. (eds.) Proceedings of the 7th International Conference on Concept Lattices and Their Applications, Sevilla, 19–21 October 2010. CEUR Workshop Proceedings, vol. 672, pp. 294–301. CEUR-WS.org (2010). http://dblp.uni-trier.de/rec/bib/conf/cla/BabinK10

7. Babin, M.A., Kuznetsov, S.O.: Computing premises of a minimal cover of functional dependencies is intractable. Discrete Appl. Math. **161**(6), 742–749 (2013). doi:10.1016/j.dam.2012.10.026. http://dblp.uni-trier.de/rec/bib/journals/dam/BabinK13

8. Bazhanov, K., Obiedkov, S.A.: Optimizations in computing the Duquenne-Guigues basis of implications. Ann. Math. Artif. Intell. **70**(1–2), 5–24 (2014). doi:10.1007/s10472-013-9353-y. http://dblp.uni-trier.de/rec/bib/journals/amai/BazhanovO14

9. Beeri, C., Bernstein, P.A.: Computational problems related to the design of normal form relational schemas. ACM Trans. Database Syst. **4**(1), 30–59 (1979). doi:10.1145/320064.320066. http://doi.acm.org/10.1145/320064.320066. http://dblp.uni-trier.de/rec/bib/journals/tods/BeeriB79

10. Borchmann, D.: Towards an error-tolerant construction of $\mathcal{E}\mathcal{L}^\perp$-ontologies from data using formal concept analysis. In: Cellier, P., Distel, F., Ganter, B. (eds.) Formal Concept Analysis, Proceedings of the 11th International Conference, ICFCA 2013, Dresden, 21–24 May 2013. Lecture Notes in Computer Science, vol. 7880, pp. 60–75. Springer, Heidelberg (2013). doi:10.1007/978-3-642-38317-5_4. http://dblp.uni-trier.de/rec/bib/conf/icfca/Borchmann13

11. Borchmann, D.: Learning terminological knowledge with high confidence from erroneous data. Ph.D. thesis, Dresden University of Technology (2014). http://dblp.uni-trier.de/rec/bib/phd/dnb/Borchmann14

12. Borchmann, D., Distel, F., Kriegel, F.: Axiomatisation of general concept inclusions from finite interpretations. J. Appl. Non-Classical Log. **26**(1), 1–46 (2016). doi:10.1080/11663081.2016.1168230. http://dblp.uni-trier.de/rec/bib/journals/jancl/BorchmannDK16

13. Davey, B.A., Priestley, H.A.: Introduction to Lattices and Order, 2nd edn. Cambridge University Press, Cambridge (2002). http://dblp.uni-trier.de/rec/bib/books/daglib/0023601

14. Distel, F.: Model-based most specific concepts in description logics with value restrictions. Tech. Rep. 08-04, Institute for theoretical computer science, TU Dresden, Dresden (2008). http://lat.inf.tu-dresden.de/research/reports.html

15. Distel, F.: Hardness of enumerating pseudo-intents in the lectic order. In: Kwuida, L., Sertkaya, B. (eds.) Formal Concept Analysis, Proceedings of the 8th International Conference, ICFCA 2010, Agadir, 15–18 March 2010. Lecture Notes in Computer Science, vol. 5986, pp. 124–137. Springer, Heidelberg (2010). doi:10.1007/978-3-642-11928-6_9. http://dblp.uni-trier.de/rec/bib/conf/icfca/Distel10

16. Distel, F.: Learning description logic knowledge bases from data using methods from formal concept analysis. Ph.D. thesis, Dresden University of Technology (2011). http://dblp.uni-trier.de/rec/bib/phd/de/Distel2011

17. Distel, F., Sertkaya, B.: On the complexity of enumerating pseudo-intents. Discrete Appl. Math. **159**(6), 450–466 (2011). doi:10.1016/j.dam.2010.12.004. http://dblp.uni-trier.de/rec/bib/journals/dam/DistelS11

18. Donini, F.M., Colucci, S., Noia, T.D., Sciascio, E.D.: A tableaux-based method for computing least common subsumers for expressive description logics. In: Boutilier, C. (ed.) IJCAI 2009, Proceedings of the 21st International Joint Conference on Artificial Intelligence, pp. 739–745, Pasadena, CA, 11–17 July 2009 (2009). http://dblp.uni-trier.de/rec/bib/conf/ijcai/DoniniCNS09

19. Facebook: Facebook (2016). https://www.facebook.com

20. Facebook for Developers: The Graph API (2016). https://developers.facebook.com/docs/graph-api

21. Ganter, B.: Two Basic Algorithms in Concept Analysis. FB4-Preprint 831, Technische Hochschule Darmstadt, Darmstadt (1984)

22. Ganter, B.: Attribute exploration with background knowledge. Theor. Comput. Sci. **217**(2), 215–233 (1999). doi:10.1016/S0304-3975(98)00271-0. http://dblp.uni-trier.de/rec/bib/journals/tcs/Ganter99

23. Ganter, B.: Two basic algorithms in concept analysis. In: Kwuida, L., Sertkaya, B. (eds.) Formal Concept Analysis, Proceedings of the 8th International Conference, ICFCA 2010, Agadir, 15–18 March 2010. Lecture Notes in Computer Science, vol. 5986, pp. 312–340. Springer, Heidelberg (2010). doi:10.1007/978-3-642-11928-6_22. http://dblp2.uni-trier.de/rec/bib/conf/icfca/Ganter10

24. Ganter, B., Wille, R.: Formal Concept Analysis - Mathematical Foundations. Springer, Heidelberg (1999). http://dblp.uni-trier.de/rec/bib/books/daglib/0095956

25. Guigues, J.L., Duquenne, V.: Famille minimale d'implications informatives résultant d'un tableau de données binaires. Math. Sci. Hum. **95**, 5–18 (1986)

26. Hitzler, P., Krötzsch, M., Rudolph, S.: Foundations of Semantic Web Technologies. Chapman and Hall/CRC Press, Boca Raton (2010). http://dblp.uni-trier.de/rec/bib/books/crc/Hitzler2010

27. Kriegel, F.: Concept Explorer FX (2010–2017). Software for Formal Concept Analysis with Description Logic Extensions. https://github.com/francesco-kriegel/conexp-fx

28. Kriegel, F.: NextClosures – parallel exploration of constrained closure operators. LTCS-Report 15–01, Chair for Automata Theory, Technische Universität Dresden (2015). http://lat.inf.tu-dresden.de/research/reports.html
29. Kriegel, F.: Axiomatization of general concept inclusions from streams of interpretations with optional error tolerance. In: Kuznetsov, S.O., Napoli, A., Rudolph, S. (eds.) Proceedings of the 5th International Workshop "What can FCA do for Artificial Intelligence"? Co-located with the European Conference on Artificial Intelligence, FCA4AI@ECAI 2016, The Hague, 30 August 2016. CEUR Workshop Proceedings, vol. 1703, pp. 9–16. CEUR-WS.org (2016). http://dblp.uni-trier.de/rec/bib/conf/ecai/Kriegel16
30. Kriegel, F.: NextClosures with constraints. In: Huchard, M., Kuznetsov, S.O. (eds.) Proceedings of the Thirteenth International Conference on Concept Lattices and Their Applications, Moscow, 18–22 July 2016. CEUR Workshop Proceedings, vol. 1624, pp. 231–243. CEUR-WS.org (2016). http://dblp2.uni-trier.de/rec/bib/conf/cla/Kriegel16
31. Kriegel, F.: Parallel attribute exploration. In: Haemmerlé, O., Stapleton, G., Faron-Zucker, C. (eds.) Graph-Based Representation and Reasoning - Proceedings of the 22nd International Conference on Conceptual Structures, ICCS 2016, Annecy, 5–7 July 2016. Lecture Notes in Computer Science, vol. 9717, pp. 91–106. Springer, Heidelberg (2016). doi:10.1007/978-3-319-40985-6_8. http://dblp2.uni-trier.de/rec/bib/conf/iccs/Kriegel16
32. Kriegel, F., Borchmann, D.: NextClosures: parallel computation of the canonical base. In: Ben Yahia, S., Konecny, J. (eds.) Proceedings of the Twelfth International Conference on Concept Lattices and Their Applications, Clermont-Ferrand, 13–16 October 2015. CEUR Workshop Proceedings, vol. 1466, pp. 181–192. CEUR-WS.org (2015). http://dblp.uni-trier.de/rec/bib/conf/cla/KriegelB15
33. Küsters, R., Molitor, R.: Computing least common subsumers in ALEN. In: Nebel, B. (ed.) Proceedings of the Seventeenth International Joint Conference on Artificial Intelligence, IJCAI 2001, pp. 219–224, Seattle, WA, 4–10 August 2001. Morgan Kaufmann, San Francisco (2001). http://dblp.uni-trier.de/rec/bib/conf/ijcai/KustersM01
34. Kuznetsov, S.O.: On computing the size of a lattice and related decision problems. Order 18(4), 313–321 (2001). doi:10.1023/A:1013970520933. http://dblp.uni-trier.de/rec/bib/journals/order/Kuznetsov01
35. Kuznetsov, S.O.: On the intractability of computing the Duquenne-Guigues base. J. UCS 10(8), 927–933 (2004). doi:10.3217/jucs-010-08-0927. http://dblp.uni-trier.de/rec/bib/journals/jucs/Kuznetsov04
36. Kuznetsov, S.O., Obiedkov, S.A.: Counting pseudo-intents and #p-completeness. In: Missaoui, R., Schmid, J. (eds.) Formal Concept Analysis, Proceedings of the 4th International Conference, ICFCA 2006, Dresden, 13–17 February 2006. Lecture Notes in Computer Science, vol. 3874, pp. 306–308. Springer, Heidelberg (2006). doi:10.1007/11671404_21. http://dblp.uni-trier.de/rec/bib/conf/icfca/KuznetsovO06
37. Kuznetsov, S.O., Obiedkov, S.A.: Some decision and counting problems of the Duquenne-Guigues basis of implications. Discrete Appl. Math. 156(11), 1994–2003 (2008). doi:10.1016/j.dam.2007.04.014. http://dblp.uni-trier.de/rec/bib/journals/dam/KuznetsovO08
38. Mac Lane, S.: Categories for the Working Mathematician. Graduate Texts in Mathematics, vol. 5, 2nd edn. Springer, New York (1978). doi:10.1007/978-1-4757-4721-8_1. http://dx.doi.org/10.1007/978-1-4757-4721-8_1
39. Maier, D.: The Theory of Relational Databases. Computer Science Press, Rockville (1983). http://dblp.uni-trier.de/rec/bib/books/cs/Maier83
40. Mantay, T.: Computing least common subsumers in expressive description logics. In: Foo, N.Y. (ed.) Advanced Topics in Artificial Intelligence, Proceedings of the 12th Australian Joint Conference on Artificial Intelligence, AI '99, Sydney, 6–10 December 1999. Lecture Notes in Computer Science, vol. 1747, pp. 218–230. Springer, Heidelberg (1999). doi:10.1007/3-540-46695-9_19. http://dblp.uni-trier.de/rec/bib/conf/ausai/Mantay99
41. Mantay, T.: A least common subsumer operation for an expressive description logic. In: Logananthoraj, R., Palm, G. (eds.) Intelligent Problem Solving, Methodologies and Approaches, Proceedings of the 13th International Conference on Industrial and Engineering

Applications of Artificial Intelligence and Expert Systems, IEA/AIE 2000, New Orleans, LA, 19–22 June 2000. Lecture Notes in Computer Science, vol. 1821, pp. 474–481. Springer, Heidelberg (2000). doi:10.1007/3-540-45049-1_57. http://dblp.uni-trier.de/rec/bib/conf/ieaaie/Mantay00

42. Obiedkov, S.A., Duquenne, V.: Attribute-incremental construction of the canonical implication basis. Ann. Math. Artif. Intell. **49**(1–4), 77–99 (2007). doi:10.1007/s10472-007-9057-2. http://dblp.uni-trier.de/rec/bib/journals/amai/ObiedkovD07

43. Sertkaya, B.: Some computational problems related to pseudo-intents. In: Ferré, S., Rudolph, S. (eds.) Formal Concept Analysis, Proceedings of the 7th International Conference, ICFCA 2009, Darmstadt, 21–24 May 2009. Lecture Notes in Computer Science, vol. 5548, pp. 130–145. Springer, Heidelberg (2009). doi:10.1007/978-3-642-01815-2_11. http://dblp.uni-trier.de/rec/bib/conf/icfca/Sertkaya09

44. Sertkaya, B.: Towards the complexity of recognizing pseudo-intents. In: Rudolph, S., Dau, F., Kuznetsov, S.O. (eds.) Conceptual Structures: Leveraging Semantic Technologies, Proceedings of the 17th International Conference on Conceptual Structures, ICCS 2009, Moscow, 26–31 July 2009. Lecture Notes in Computer Science, vol. 5662, pp. 284–292. Springer, Heidelberg (2009). doi:10.1007/978-3-642-03079-6_22. http://dblp.uni-trier.de/rec/bib/conf/iccs/Sertkaya09

45. Stumme, G.: Attribute exploration with background implications and exceptions. In: Studies in Classification, Data Analysis, and Knowledge Organization, pp. 457–469. Springer, Berlin/Heidelberg (1996). doi:10.1016/S0304-3975(98)00271-0. http://dblp.uni-trier.de/rec/bib/journals/tcs/Ganter99

46. (W3C), W.W.W.C.: Owl 2 web ontology language document overview, 2nd edn. (2012). https://www.w3.org/TR/owl2-overview/

47. Wild, M.: Computations with finite closure systems and implications. In: Du, D., Li, M. (eds.) Computing and Combinatorics, Proceedings of the First Annual International Conference, COCOON '95, Xi'an, 24–26 August 1995. Lecture Notes in Computer Science, vol. 959, pp. 111–120. Springer, Heidelberg (1995). doi:10.1007/BFb0030825. http://dx.doi.org/10.1007/BFb0030825. http://dblp.uni-trier.de/rec/bib/conf/cocoon/Wild95

48. Wille, R.: Restructuring Lattice Theory: An Approach Based on Hierarchies of Concepts, pp. 445–470. Springer, Dordrecht (1982). doi:10.1007/978-94-009-7798-3_15. http://dx.doi.org/10.1007/978-94-009-7798-3_15

Formal Concept Analysis of Attributed Networks

Henry Soldano, Guillaume Santini, and Dominique Bouthinon

1 Introduction

Our purpose is to investigate and analyze attributed graphs. In this article we discuss how recent extensions of Formal Concept Analysis apply to this problem. We consider undirected graphs $G(O, E)$ where E is the edge set, and O the vertex set. The vertices are labeled by a description in an attribute pattern language L with a lattice structure, typically $L = 2^I$ where I is a set of binary attributes. Note that we may consider such an attributed graph both as a graph whose vertices are labeled with subsets of I and as set of objects each described by such a subset of I and that may be related together by edges.

The former view leads to consider the methodology used to investigate graphs, in particular social and complex networks. Most of the work in this area consider unlabeled networks and is concerned by what may be said about the topological structure of the network. A large set of measures have been proposed to analyze these networks, and two main ways have been proposed to extract interesting subgraphs. The first way consider the network as made of a *core*, i.e., a dense subgraph whose vertices are highly connected, together with its *periphery*, made of vertices highly connected to the core, but poorly connected between them [6].

H. Soldano (✉)
Université Paris 13, Sorbonne Paris Cité, L.I.P.N UMR-CNRS 7030, F-93430, Villetaneuse, France

Museum National d'Histoire Naturelle, ISYEB - UMR 7205 CNRS MNHN UPMC EPHE, F-75005, Paris, France
e-mail: henry.soldano@lipn.univ-paris13.fr

G. Santini • D. Bouthinon
Université Paris 13, Sorbonne Paris Cité, L.I.P.N UMR-CNRS 7030, F-93430, Villetaneuse, France
e-mail: guillaume.santini@lipn.univ-paris13.fr; dominique.bouthinon@lipn.univ-paris13.fr

© Springer International Publishing AG 2017
R. Missaoui et al. (eds.), *Formal Concept Analysis of Social Networks*,
Lecture Notes in Social Networks, DOI 10.1007/978-3-319-64167-6_6

143

The second way considers the network as made of a number of dense subnetworks, called *communities* whose vertices are highly connected within the community and poorly connected to vertices of other communities [8]. Finally, the two views may be combined, for instance, by considering the network as made of communities each having some core/periphery structure [16].

Regarding the notion of core, there have been various ways to define it, starting from the k-core of a network which is the greatest subnetwork whose vertices all have degree at least k in the subnetwork [17]. By changing the topological property, but keeping this idea of a greatest subnetwork whose vertices share the property within the network, we obtain various core definitions [3]. A core may also be defined as the greatest subnetwork made of a subset of a family of small, connected subnetworks. The simplest example is the k-clique core that is only made of k-cliques. When $k = 3$, the core is made of triangles which are known to be an important substructure in social networks analysis [29].

Concerning the idea of communities, it has been extensively investigated mostly as an optimization problem: how to optimally partition the network in subnetworks maximizing some measure. A second view of communities derives from some strong structural property that has to be satisfied within a community. We will further call them *structural communities*, or simply communities, as we only consider these kind of communities in the remaining part of this article. The main example is the k-community approach that divides a k-clique core (as defined above) in connected subnetworks each satisfying a stronger property (see below) [13] .

From the first point of view, adding attributes to the vertices means that each attribute pattern induces a subgraph whose vertices satisfy the pattern. Each such subgraph could then be investigated, extracting its core and communities. The question is then how to summarize and select relevant information from such a set of results.

The second view considers the attributed graph first as a table representing a set of objects described by attributes, and then considers that edges may relate objects. This leads to the use of standard methodology of data analysis by adapting it to dealing with topological information. The whole purpose of this article is to discuss how Formal Concept Analysis, which was originally concerned with data tables, may be extended in order to take into account the topological information. The main idea we propose here is to consider as parameters the notion of cores and structural communities relevant to the data to analyze and adapt accordingly the FCA methodology.

Regarding the reduction of a graph to its core, this may be obtained by defining an interior operator p on the vertex powerset 2^O. This approach, based on a previous work on abstraction in Formal Concept Analysis [24], produces *abstract closed patterns* structured in an *abstract concept lattice* together with a basis of *abstract implications* written $\Box q \rightarrow \Box w$. All concept extents are then images of the interior operator. When considering attributed networks, and given some core definition, such an interior operator reduces a vertex subset e to the vertices forming the core of the subgraph $G(e)$ induced by e [23]. It is then called a *graph abstraction* operator. Fig. 1 represents an attributed graph G, the subgraph induced by pattern a (in plain

Fig. 1 The pattern *a* subgraph is displayed with *plain lines*, the corresponding 3-clique abstract subgraph is displayed in *blue lines*. The associated abstract closed pattern is *ab*

Fig. 2 The DMKD,IDArev pattern subgraph in the DBLP co-authoring experiment. The *red vertices* and *edges* represent the subgraph induced by the degree ≥ 4 abstract extension

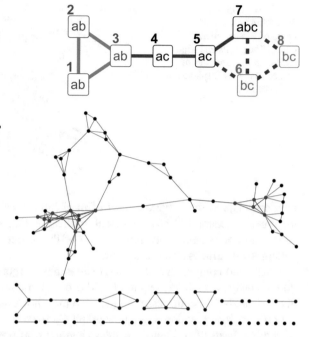

lines), and the 3-clique core of this subgraph, i.e., its *abstract subgraph* (in blue lines). Note that all vertices of the core share attribute *b*. This means that *ab* is an abstract closed pattern and that the abstract implication $\Box a \rightarrow \Box ab$ holds, i.e., any triangle in *G* whose vertices share *a* also share *b*.

Recent works in attributed graph mining are interested in searching for local patterns made of a constraint on a subset of attributes together with a density constraint on a vertex subset, and this using various notions of maximality [11, 18]. In a companion article [21], we have defined *local closed patterns* corresponding to maximal attribute patterns each associated with one dense subgraph, allowing to extract *local implications*, particular to specific dense groups of objects. For that purpose Formal Concept Analysis (FCA) had to be extended in order to take into account this notion of locality.

The simplest example is obtained by considering a subgraph made of various connected components, and associating to each connected component a local closed pattern, i.e., the most specific pattern shared by the vertices of this connected component. More generally, local closed patterns may be associated with the connected components of abstract subgraphs. The family of such connected components forms a partial order called a *cc-confluence* while the corresponding *local concepts* have a weaker structure called a *pre-confluence*. As an example, in Fig. 2 we display a pattern subgraph extracted from a DBLP co-authoring network labeled by journal and conference names, together with its abstract subgraph (in bold and red vertices and lines) when considering a 4-core abstraction. The abstract subgraph has two connected components, i.e., two structural communities of scientists. Again we

Fig. 3 The original
Friendship network of a
group of West Scotland
pupils. The pupils and edges
forming 3-communities of
size at least 4 are displayed in
various colors

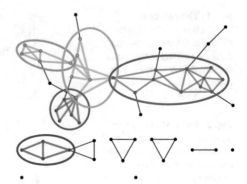

may associate to this structure a set of implications, called *local implications*. In the previous example we found a local implication $\square_i q \rightarrow \square_i w$ stating that in the connected component containing vertex i of the degree ≥ 4 abstract subgraph of pattern q, all vertices also share pattern w.

Connected components of abstract subgraphs as represented in *cc*-confluences do not always completely capture the idea of communities as considered in social network analysis. As discussed in [21], we may, however, enlarge the local closed patterns approach by deriving a new graph G_T from G whose vertex set T is a set of vertex subsets of G. Figure 3 displays a graph whose vertices represent pupils of a school in the West of Scotland, whose edges represent friendship relations, and whose vertex attributes concern substance use and sporting activity.[1] As a running example we consider the subgraph induced by the empty pattern, i.e., the whole graph. By applying a 3-clique graph abstraction restricted to connected components containing at least 4 vertices,[2] we obtain a subgraph made of the bold and colored edges and vertices. This abstract subgraph is made of two connected components, therefore leading to two local concepts. However, the largest connected component is clearly made of distinct dense parts, i.e., *communities*, we would like to consider when defining local closed patterns. Fortunately, when considering k-communities [13] we can solve this problem by applying the cc-confluence approach to a new graph derived from the original graph. More precisely, a k-community is a vertex subset in a graph G that corresponds to a connected component in a derived graph G_T. The vertices of G_T are k-cliques in G and an edge relates two vertices whenever the corresponding k-cliques share $k - 1$ vertices in G. Each colored subgraph in Fig. 3 defines such a 3-community.

The last task is to define interestingness measures to rank abstract or local patterns and implications. Regarding patterns, we will search for patterns whose abstract (resp. local) subgraph is a large part of the whole pattern subgraph, i.e., which preserves the core (resp. communities) definitions. The corresponding measure is called *specificity*. Regarding the abstract and local implications, we search for

---

[1] http://www.stats.ox.ac.uk/~snijders/siena/s50_data.htm.

[2] We further call the 3-*clique and cc-$\geq$4* abstraction.

implications which are informative, i.e., which did not hold as standard implications and therefore bring some new information about our data. The corresponding measure is called *Informativity*. A preliminary discussion of both measures was presented in [25]. We propose here definitions of *specificity* and *informativity* both at the abstract and local level and experiment them on two real attributed networks.

Section 2 describes the attributed graphs used in our experiments. Section 3 presents abstract concept lattices, abstract implications, and graph abstractions together with associated interestingness measures. Section 4 defines local concept pre-confluences, related local implications, *cc*-confluences, and interestingness measures. In Sect. 5 we show how we extract the set of 3-communities associated with pattern subgraphs by using derived *cc*-confluences, and we display the local concept ordering of the attributed network of teenage friendship displayed in Fig. 3. In Sect. 6 we briefly discuss the implementation used in our experiments.

2 Datasets

In this section we will consider experiments with two datasets. In both cases the data are described as a graph $G = (O, E)$. Vertices of this graph are have labels from 2^I, where I is a set of items, i.e., binary attributes. Since objects are not always described by binary attributes, the binarization preprocessing is described when necessary.

2.1 Teenage Friends and Lifestyle Study

The dataset is denoted by *s50-1* and is a standard attributed graph dataset.[3] It represents 148 friendship relations between 50 pupils of a school in the West of Scotland, and labels concern the substances used (tobacco, cannabis, and alcohol) and sporting activity. The values of the corresponding variables are ordered. The binarization process consists in defining variables representing the value intervals. T stands for Tobacco consumption and has values 1 (no smoking), 2 (occasional), and 3 (regular). C stands for cannabis consumption and has values 1 (never tries) to 4, D stands for alcohol consumption and has values 1 (does not drink) to 5, and S stands for sporting activity and has two values 1 (occasional) and (2) regular. Binary attributes represent intervals, for instance, C234 means that the value of C is at least 2 and therefore represents the interval [2, 4]. In Table 1 we present the binary attributes we have defined. Attribute subsets represent intersection of intervals. For example pattern {D123, D2345} requires that the value of D lies within the interval $[1, 3] \cap [2, 5] = [2, 3]$. Note that, for the sake of simplicity we do not distinguish the two highest values of attributes T, C, and D.

---

[3]http://www.stats.ox.ac.uk/~snijders/siena/s50_data.htm.

Table 1 The binary attributes used to label the vertices in the Teenage Friendship network

Tobacco	Cannabis	Alcohol	Sport
T1, T23	C1, C12, C234, C34	D1, D12, D123, D2345, D345, D45	S1, S2

2.2 A DBLP Dataset

This is the DBLP dataset as described in [4]. There are 45,131 vertices, 228,188 edges, and 555 connected components. Vertices are authors that have published at least one paper in one among 29 journals or conferences of the Database and Datamining communities[4] during the 1/1990–2/2011 period. An edge links two authors whenever they are coauthors of at least one article. The conferences are clustered in three clusters: DB (databases), DM (data mining), and AI (artificial intelligence) according to a conference ranking site categorization.[5] The binary attributes are the journal and conference names together with the three clusters. An attribute has value 1 if the author has published in the corresponding journal or conference or cluster.

3 Abstract Closed Patterns in Attributed Networks

3.1 Closed Patterns

In this section we introduce the necessary definitions and terminology we use in the article. Note that the terminology is somewhat non-standard in FCA. Indeed, as we need to interleave interior operators with extensional and intensional operators, the standard X'' notation to represent closed elements is not so convenient, so we rather denote, respectively, by ext and int the extensional and intensional operators. Furthermore, in this introductory paragraph we relate FCA to closed pattern mining.

A standard pattern mining procedure consists in considering the set of occurrences of patterns, belonging to some pattern language L with a lattice structure,[6] within an object set O (see, for instance, [5]). This language is partially ordered following a general-to-specific ordering, and each object o is described as a

---

[4]Conferences: KDD, ICDM, ECML/PKDD, PAKDD, SIAM DM, AAAI, ICML, IJCAI, IDA, DASFAA, VLDB, CIKM, SIGMOD, PODS, ICDE, EDBT, ICDT, SAC ; Journals: IEEE TKDE, DAMI, IEEE Int. Sys., SIGKDD Exp., Comm. ACM, IDA J., KAIS, SADM, PVLDB, VLDB J., ACM TKDD.

[5]http://webdocs.cs.ualberta.ca/~zaiane/htmldocs/ConfRanking.html.. DB = {VLDB, SIGMOD, PODS, ICDE, ICDT, EDBT, DASFAA, CIKM}; DM= {SIGKDD Explorations, ICDM, PAKDD, ECML/PKDD, SDM}; AI= {IJCAI, AAAI, ICML, ECML/PKDD}.

[6]We recall that in a lattice any pair of elements (x, y) has a greatest lower bound $x \wedge y$ (or *meet*) and a least upper bound (or *join*) $x \vee y$.

particular pattern $d(o)$. A pattern q occurs in object o whenever $d(o)$ is more specific (i.e., larger) than q. The set of occurrences $\mathrm{ext}(q)$ of a pattern q is called its *extension*. An intension function $\mathrm{int}(e)$ returns the most specific pattern associated with the extension e. This means that we relate a pattern q to the most specific pattern with same extension by applying the *closure operator* $\mathrm{int} \circ \mathrm{ext}$ to q. $\mathrm{int} \circ \mathrm{ext}(q)$ is then called a closed pattern. The pattern language L typically is 2^I where I is a set of binary attributes (aka items). With no loss of generality we will further use the powerset 2^I as a pattern language while what follows also applies to wider languages as pattern structures[10]. When $L = 2^I$ the closure operator on patterns then simply intersects the object descriptions of the extension of the entry pattern. This means that when considering patterns with same extension as equivalent, closed patterns are the representatives of the equivalences classes. Such a class has therefore a maximum but also minimal elements, called *minimal generators*. When the patterns belong to 2^I, the min–max basis of implications[14] is defined as follows:

$$m = \{g \to f \backslash g \mid f \text{ is a closed pattern}, g \text{ is a generator}, f \neq g, \mathrm{ext}(g) = \mathrm{ext}(f)\}$$

This basis represents all the implications $t \to t'$ that hold on O, i.e., such that $\mathrm{ext}(t) \subseteq \mathrm{ext}(t')$. This precisely means that all these implications may be derived from the min–max basis. Obviously all non-trivial implications, i.e., implications such that $t' \not\subseteq t$, may be inferred from an implication $l \to r$ of m where $l \subseteq t$ and $r \cup l \supseteq t'$.

Finally, note that the enumeration of closed patterns is in general restricted to frequent patterns, i.e., patterns whose extension is larger than some threshold. In FCA, such a constraint leads to iceberg lattices [27].

3.2 Abstract Closed Patterns

We summarize here how abstraction is applied in FCA by constraining the extensional space. We first recall the definitions of closure operators and *interior operators*, the latter being further used to restrict the pattern extensions to be *abstract extensions*. In what follows all ordered sets are finite, and in particular any topped meet-semilattice (resp. pointed join-semilattice) is a lattice.

Definition 1 Let U be an ordered set and $f : U \to U$ a self map such that for any $x, y \in U, f$ is monotone, i.e., $x \leq y$ implies $f(x) \leq f(y)$ and idempotent, i.e., $f(f(x)) = f(x)$, then:

- If f is extensive, i.e., $f(x) \geq x, f$ is called a closure operator
- If f is intensive, i.e., $f(x) \leq x, f$ is called a dual closure operator, an interior operator, or also a projection.

In the first case, an element such that $x = f(x)$ is called a closed element.

Ranges of interior operators on lattices are called *abstractions* and are characterized by the following Proposition:

Proposition 1 (see [24]) *A subset A of X $= 2^O$ is the range p[X] of some interior operator p on X, if and only if for any elements x, y in A, their join $x \cup y$ also belongs to A and A contains the empty set. The interior operator is related to its range as follows:*

$$p(x) = \sup_{\{a \in A | a \subseteq x\}} a.$$

Let then p be the interior operator associated with some abstraction A, $p(x)$ be the greatest element of A contained in x. Closed pattern analysis has been recently extended to *abstract* closed pattern analysis by noticing that applying an interior operator on the extensional space 2^O we obtain again a closure operator on the pattern language 2^I [15, 24]:

Proposition 2 *Let X $= 2^O$ and L $= 2^I$, p be an interior operator on 2^O, and A $= p[X]$ be the associated abstraction, we have that (int, p $\circ$ ext) is a Galois connection on (A, L), i.e.,:*
  f $=$ int $\circ$ p $\circ$ ext is a closure operator on L,
  The *abstract extension* of pattern q is defined as $p \circ \text{ext}(q)$. A new equivalence relation is then defined such that $q \equiv_A w$ whenever $p \circ \text{ext}(q) = p \circ \text{ext}(w)$, each equivalence class of which corresponds to some abstract extension in A. There is then a unique *abstract support closed* pattern, i.e., a most specific pattern among all patterns sharing the same abstract extension, which is obtained as $f(q) = \text{int} \circ p \circ \text{ext}(q)$. $f(q)$ is then called an *abstract closed pattern*. This leads to the definition of abstract concepts organized in a concept lattice:

Corollary 1 ([24]) *The set of (abstract extension, abstract closed pattern) pairs (e $=$ ext(c), c $=$ int(e)), ordered following A, is a lattice called an abstract concept lattice.*
  Note that, as p is monotone, whenever $\text{ext}(q) \subseteq \text{ext}(w)$, i.e., $q \rightarrow w$ is valid, we also have $\text{ext}_A(p) = p \circ \text{ext}(q) \subseteq \text{ext}_A(w) = p \circ \text{ext}(w)$. The latter inclusion states the validity of an *abstract implication* we will rewrite as $\square^A q \rightarrow \square^A w$.
  This way we obtain *abstract min–max basis* with the same definition as earlier in this section except that ext_A replaces ext and therefore abstract implications relate minimal elements (i.e., A-generators) to maximal element (the abstract closed pattern, or A-closed pattern) of the same abstract equivalence class. We have then the following definition:

Definition 2 The *abstract min–max basis* m_A of valid abstract implications is defined as
  $m_A = \{\square^A g \rightarrow \square^A f \backslash g \mid f$ is an A-closed pattern, g is a A-generator, $f \neq g$, $\text{ext}_A(g) = \text{ext}_A(f)\}$.
  In the same way as in the standard min–max basis case, all implications $\square^A t \rightarrow \square^A t'$ that hold on A, i.e., such that $\text{ext}_A(t) \subseteq \text{ext}_A(t')$, may be inferred from the abstract min–max basis.

3.3   Graph Abstractions

These ideas have been applied to attributed graphs by defining graph abstractions [23]. The set of objects O is then the set of vertices of a graph $G = (O, E)$, and each vertex o is labeled by an attribute pattern $d(o) \in 2^I$.

A graph abstraction is an abstraction of 2^O defined through a characteristic property P such that $P(x, e)$ expresses some minimal connectivity requirement of the vertex x within the subgraph G_e induced by some vertex subset e.

Proposition 3 *Let P be such that*

- *$P(x, e)$ implies $x \in e$ and*
- *$e \subseteq e'$ and $P(x, e)$ implies $P(x, e')$,*

and let q be a mapping defined by $q(e) = \{x \in e | P(x, e)\}$, then the mapping p defined by $p(e) = \text{fixedpoint}(q, e)$ is an interior operator on 2^O.

Consider a subgraph G_e, where $p(e)$ represents the greatest vertex subset of e inducing a subgraph whose vertices all satisfy the associated characteristic property. This subgraph $G_{p(e)}$ will be further called the *abstract subgraph* of G_e. We give hereunder examples of graph abstractions, defined through their characteristic property and exemplified in Fig. 4.

1. degree $\geq k$. The degree $\geq k$-abstract subgraph of a graph is its k-core [17].
2. k-club $\geq s$: x has to belong to at least one k-club of size at least s in G_e. This is a relaxation of the notion of clique[1]: a k-club is a subset c of vertices such that there is a path of length $\leq k$ between any pair of vertices in G_c. A triangle, a 3-clique, is a 1-club of size 3 (Fig. 4a). Figure 4b represents a 2-club of size 6 and therefore a 2-club≥ 6 abstract group.
3. nearStar(k, d): x has to have degree at least k or there must be a path of length at most d between x and some y with degree at least k. For instance, the simplest nearStar(8, 1) abstract group is a central node connected with eight nodes. Such an abstraction is useful when we want the abstraction to preserve hubs [2] (i.e., high degree vertices) together with their (low degree) neighbors (see Fig. 4c).
4. $cc \geq s$: x has to belong to a connected component of size at least s in G_e (see Fig. 4d).

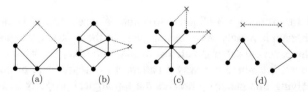

(a)   (b)   (c)   (d)

Fig. 4 Graph abstractions corresponding to various vertex characteristic properties. In each graph *plain circles* and *plain lines* form the abstract subgraph, *crosses* and *dotted lines* represent the vertices and edges out of the abstract subgraph. (**a**) x has to belong to a 3-clique, (**b**) x has to belong to a 2-club of size at least 6, (**c**) x has to be connected to a vertex y such that the degree of y is at least 6, i.e., to a nearstar(6,1), (**d**) x has to belong to a connected component whose size is at least 3

Finally, it is interesting to note that we can combine two (or more) abstractions A_1 and A_2 in two ways, defining a new composite abstraction either stronger or weaker than both A_1 and A_2. For instance, we may want to consider an abstract subgraph where vertices both have a degree larger than some k and belong to a connected component exceeding a minimal size s. On the contrary, we may want an abstract subgraph such that at least one of the two characteristic properties is satisfied by all the vertices. This would be the case, for instance, if we want to keep both vertices that have a degree larger than, say 10, and vertices in a star, i.e., connected to a hub which degree is at least 50. The following proposition states that we can freely combine abstractions in both directions.

Proposition 4 *Let P_1 and P_2 two characteristic properties of abstractions defined on the same object set O, and let $P_1 \wedge P_2$ and $P_1 \vee P_2$ be defined as follows:*

- $P_1 \wedge P_2(x, e) = P_1(x, e) \wedge P_2(x, e)$
- $P_1 \vee P_2(x, e) = P_1(x, e) \vee P_2(x, e)$

Both $P_1 \wedge P_2$ and $P_1 \vee P_2$ are characteristic properties of abstractions.

3.4 Interestingness Measures on Abstract Patterns and Implications

3.4.1 Specificity of Abstract Patterns

We are now interested in measuring knowledge brought by abstract closed patterns and abstract implications [25]. For that purpose we first generalize hereunder the *structural correlation* measure introduced by A. Silva and co-authors [19], originally introduced to relate a subgraph to its content in terms of quasi-cliques and rename it as *specificity*.

Definition 3 Let q be a pattern, A an abstraction of some powerset of objects O, the specificity of q with respect to A is defined as:

$$S_A(q) = \frac{|\operatorname{ext}_A(q)|}{|\operatorname{ext}(q)|}$$

Consider, for instance, a 3-clique abstraction. Whenever $S_A(q)$ is close to 1, the pattern q subgraph is mainly made of triangles. To the contrary, whenever $S_A(q)$ is close to 0, the pattern q subgraph almost displays no triangles, which means quite isolated vertices. We relate this way a pattern q to the measure of how selecting vertices satisfying this pattern preserves the topological property associated with the abstraction.

Example 1 Figure 1 displays a graph each vertex of which is described by an itemset. We observe then that:

- $\text{ext}(a) = e = \{1, 2, 3, 4, 5, 7\}$ induces the subgraph $G(e)$ (blue+black).
- $\text{ext}_A(a) = \{1, 2, 3\}$ as $4, 5, 7$ do not belong to any 3-clique in $G(e)$.
- $\text{int} \circ \text{ext}_A(a) = ab \cap ab \cap ab = ab$ is an abstract closed pattern.
- $S_A(a) = 1/2, S_A(ab) = 3/4$.

Note that among the patterns of some equivalence class of $\equiv_A$ the abstract closed pattern c has maximal specificity:

For any t, if $\text{ext}_A(t) = \text{ext}_A(c)$ then $S_A(c) \geq S_A(t)$

3.4.2 Informativity of Abstract Implications

Apart from measuring through specificity what is specific to the pattern in its abstract view, we are also interested when considering abstract implications in how informative they are. For that purpose we consider abstract implications whose left and right patterns are equivalent in the abstract space A, i.e., have same abstract extension, as in the min–max abstract implication basis defined above. Whenever these patterns are also equivalent in the original space 2^O intuitively the implication is uninformative. Assume, for instance, that $a \rightarrow abc$ is valid, then validity of the abstract implication with same left and right members does not bring any new information. On the contrary, assume that $\Box^A a \rightarrow \Box^A abc$ is valid while $a \rightarrow abc$ has only confidence 0.5, i.e., $\text{ext}(abc) = 0.5 * \text{ext}(a)$, then clearly the abstract implication brings some information.

Definition 4 Let q be a pattern, A an abstraction of 2^O, the *informativity* of the valid implication $r : \Box^A q \rightarrow \Box^A w$ is defined as:

$$I_A(r) = 1 - \frac{|\text{ext}(qw)|}{|\text{ext}(q)|}$$

Informativity has a range between 0 and 1 and estimates the probability of not having w whenever we have q in graph G. This quantity has value 0 whenever $q \rightarrow w$ holds and has limit 1 whenever $|\text{ext}(qw)|$ approaches 0, i.e., restricting the extension of patterns to elements of A concentrates the extension of q to the very few sharing also w.

Intuitively, the informativity of an abstract implication measures what we discovered when we observed that q and qw share the same abstract support.

Example 2 Following Example 1 illustrated in Fig. 1 consider the abstract implication $r : \Box a \rightarrow \Box ab$. This abstract implication has the following semantics: "a 3-clique of G whose vertices share pattern a also share pattern b," and its informativity is therefore $I_A(r) = 1 - 1/2 = 0.5$.

Note that implications of the abstract min–max basis which relate minimal elements g of an abstract equivalence class to the corresponding abstract closed pattern c have maximal informativity:

Let $g \equiv_A t \equiv_A t' \equiv_A c$ then $I_A(\Box^A g \rightarrow \Box^A c) \geq I_A(\Box^A t \rightarrow \Box^A t')$.

Table 2 Top-15 abstract closed patterns in the Teenage Friendship network ranked according to their 3-clique and cc≥4 specificity

N°	$\Box^A c$	$|ext^A(c)|$	$|ext(c)|$	$S_A(c)$
1	$\Box^A$D45-C34	5	7	0.714
2	$\Box^A$C12	27	42	0.643
3	$\Box^A\emptyset$	32	50	0.64
4	$\Box^A$C12-T1	21	36	0.583
5	$\Box^A$D45	9	17	0.529
6	$\Box^A$D345	15	29	0.517
7	$\Box^A$C1-T1	17	33	0.515
8	$\Box^A$D123-C12-T1	15	30	0.5
9	$\Box^A$D345-C12	10	21	0.476
10	$\Box^A$D2345	21	45	0.467
11	$\Box^A$C12-S2	15	33	0.455
12	$\Box^A$D45-C12-S2	4	9	0.444
13	$\Box^A$D2345-C12	16	37	0.432
14	$\Box^A$S2	16	37	0.432
15	$\Box^A$D123-C1-T1	11	27	0.407

3.5 Experiments

Some experiments on the two datasets described in Sect. 2 have been performed and discussed in [23]. We discuss hereunder new experiments in particular regarding the interestingness measures.

We firs consider the Teenage Friendship network s50. Among the 50 pupils 38 belong to triangles (i.e., 3-cliques). As there are no isolated triangles, the abstract subgraph when considering only connected components with size at least 4 is reduced to 32 pupils. The corresponding abstraction is therefore *3-clique and* $cc \geq 4$.

Table 2 displays the top 15 patterns according to the corresponding specificity. We observe that specificity is clearly non-monotonic with respect to abstract extension size and that among top patterns we find both small (and therefore general) and large patterns. We further discuss the pattern with highest specificity D45-C34, which corresponds to pupils with high alcohol and cannabis consumption, in Sect. 5 where we search for communities.

Table 3 displays the top 15 abstract implications according to the *3-clique and* $cc \geq 4$ abstract informativity. Again, informativity is clearly nonmonotonic with respect to extension size. The first and third implications have the same abstract pattern as their rightmost member: it concerns the same abstract subgraph, whose pupils have in common the behavior D45-C12-S2, but is obtained either by reducing the pattern D345-S2 subgraph or the pattern D45-S2 subgraph. Obviously the former subgraph corresponds to a higher informativity as it includes the latter subgraph. As a matter of fact, the third implication is *redundant* with the first one and could be removed with no information loss. This leads to reduce the abstract min–max basis by eliminating all such redundant rules. We will apply such an idea

Table 3 Top-15 abstract implications in the Teenage Friendship network ranked according to their 3-clique and cc$\geq$4 informativity

N°	$\Box^A g \rightarrow \Box^A c$	$|ext(c)|$	$|ext(g)|$	$I_A(r)$
1	$\Box^A$D345-S2 $\rightarrow$ $\Box^A$D45-C12-S2	9	22	0.591
2	$\Box^A$C234 $\rightarrow$ $\Box^A$D45-C34	7	14	0.5
3	$\Box^A$D45-S2 $\rightarrow$ $\Box^A$D45-C12-S2	9	13	0.308
4	$\Box^A$T1-D345 $\rightarrow$ $\Box^A$D345-C1-T1	14	18	0.222
5	$\Box^A$D23 $\rightarrow$ $\Box^A$D23-C1-T1	23	28	0.179
6	$\Box^A$C1-D345 $\rightarrow$ $\Box^A$D345-C1-T1	14	17	0.176
7	$\Box^A$C34 $\rightarrow$ $\Box^A$D45-C34	7	8	0.125
8	$\Box^A$D2345-S2 $\rightarrow$ $\Box^A$D2345-C12-S2	29	33	0.121
9	$\Box^A$T1-D2345 $\rightarrow$ $\Box^A$D2345-C1-T1	29	33	0.121
10	$\Box^A$C1-D2345-S2 $\rightarrow$ $\Box^A$D2345-C1-S2-T1	22	25	0.12
11	$\Box^A$C1-S2- $\rightarrow$ $\Box^A$C1-S2-T1	25	28	0.107
12	$\Box^A$C12-D45 $\rightarrow$ $\Box^A$D45-C12-S2	9	10	0.1
13	$\Box^A$D12 $\rightarrow$ $\Box^A$D12-C1-T1	19	21	0.0952
14	$\Box^A$C1-D2345 $\rightarrow$ $\Box^A$D2345-C1-T1	29	32	0.0938
15	$\Box^A$D123 $\rightarrow$ $\Box^A$D123-C12-T1	30	33	0.0909

when defining a basis for local implications in Sect. 4.2. The second implication in this ranking concerns pattern D45-C34, mentioned as the highest specificity pattern. The implication states that the *3-clique and cc* $\geq$ 4 subgraph of pupils that have pattern C234, which corresponds to a medium-to-high cannabis consumption behavior, selects pupils with the D45-C34 pattern, i.e., those who have both high alcohol and cannabis consumption behavior. When applying no abstraction, the implication only holds on 7 among the 14 pupils that have pattern C234, which results in informativity $1 - 7/14 = 0.5$.

We discuss now some new details on experiments performed on the DBLP dataset. The experiment consisted in applying a degree $\geq k$ abstraction with increasing k-values and we focused in abstract patterns obtained with $k = 16$, which corresponds to a very strong abstraction: in an abstract extension each author is required to have 16 co-authors within the abstract extension. We obtained few abstract closed patterns and in particular the abstract closed pattern VLDBJ, ICDE, SIGMOD, VLDB and the related abstract implication $\Box$ VLDBJ $\rightarrow$ $\Box$ ICDE, SIGMOD, VLDB. Both the abstract closed pattern and its abstract minimal generator VLDBJ have an abstract extension of 38 among the 1276 VLDBJ authors in the dataset. The implication states that a dense group of co-authors that have published in the Very Large Database Journal also have published in several database conferences. In Fig. 5 we present the corresponding subgraph. Such a very dense co-authoring subgraph within the VLDBJ subgraph is somewhat unexpected. Its abstract specificity $38/1276 \approx 0.085$ is low, but still higher than the 0 value we could expect from such a high abstraction level. The abstract implication has a high informativity about ≈ 0.65 coming from the fact that among the 1276 authors who published in VLDBJ journal only 441 did publish in all the conferences ICDE, SIGMOD, and VLDB.

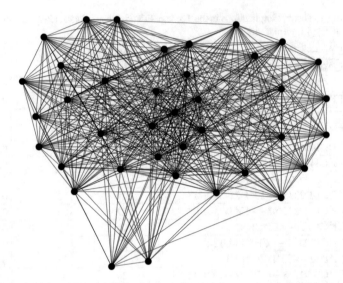

Fig. 5 The subgraph obtained when applying the degree $\geq$ 16 abstraction to the VLDBJ subgraph in the DBLP co-authoring experiment

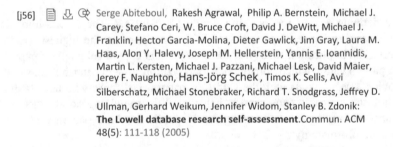

[j56] 📄 ⤓ ⊕ Serge Abiteboul, Rakesh Agrawal, Philip A. Bernstein, Michael J. Carey, Stefano Ceri, W. Bruce Croft, David J. DeWitt, Michael J. Franklin, Hector Garcia-Molina, Dieter Gawlick, Jim Gray, Laura M. Haas, Alon Y. Halevy, Joseph M. Hellerstein, Yannis E. Ioannidis, Martin L. Kersten, Michael J. Pazzani, Michael Lesk, David Maier, Jerey F. Naughton, Hans-Jörg Schek, Timos K. Sellis, Avi Silberschatz, Michael Stonebraker, Richard T. Snodgrass, Jeffrey D. Ullman, Gerhard Weikum, Jennifer Widom, Stanley B. Zdonik: **The Lowell database research self-assessment**.Commun. ACM 48(5): 111-118 (2005)

Fig. 6 An example of reference with many authors that leads to a high degree subnetwork

We made then some investigations in the DBLP repository, focussing of the 38 authors of the abstract extension, and found an article whose reference is given in Fig. 6 and whose abstract begins as follows:

> A group of senior database researchers gathers every few years to assess the state of database research …

In some sense the explanation of the pattern we discovered is straightforward. However, the whole purpose of pattern mining is to find unexpected patterns, hidden within large datasets, and interpret them in order to acquire some new knowledge. It is exactly what happens here: we were not aware of these regular meetings of senior database researchers, and we learned something new, though, of course, this knowledge is clearly widely known within the database community.

When considering a weaker abstraction, namely here a degree $\geq$ 4 abstraction, we obtain more abstract closed patterns sometimes made of several connected components. Figure 2 in Sect. 1 represents the DMKD, IDArev pattern subgraph together with the subgraph induced by the abstract extension of the pattern. This abstract subgraph is made of two connected components, the one in the right part of the figure is made of ten vertices and we are then interested in knowing whether there is some more specific pattern than the abstract closed pattern DMKD, IDArev, which would be shared by this connected component. Answering such questions means mining at a local level the attributed graph, and this is the subject of the next section.

4 Local Closed Patterns in Attributed Networks

Given some attribute pattern, we are now interested in extracting *local support closed patterns*, i.e., maximal attribute patterns each associated with one dense subgraph, so allowing to extract *local implications* particular to specific dense groups of objects. Recently FCA has been extended to *local closed patterns*: they are obtained by applying a set of *local closure operators* [22]. In the graph case, this means that from the extension of some (closed) pattern c, various dense extensions, called *local extensions* are extracted each associated with a *local closed pattern*, i.e., the most specific pattern l common to the elements of the local extension. Again we obtain a set of *local implications* corresponding to inclusion of local extensions, but now such an implication is only valid in the vicinity of some dense group of vertices.

In [21] we introduced locality in the closure framework with the main motivation of investigating local patterns in attributed graphs. For that purpose we have first to define pre-confluences and confluences which are structures weaker than lattices that have been investigated in FCA [21, 23]. Confluences, in particular, are close to but different from confluent families as defined in [5]. We further denote by E^x the up sets $\{y \in E | y \geq x\}$ of an ordered set E, by E_x its down sets $\{y \in E | y \leq x\}$, and by $\min(E)$ the set of its minimal elements.

First note that ordered sets we consider are all finite. We define a pre-confluence as a finite ordered structure that generalizes the (finite) lattice structure:

Definition 5 A finite ordered set F is a pre-confluence if and only if for any $m \in \min(F)$, $F^m = \{x \in F \mid x \geq m\}$ is a lattice.

A consequence of this definition is that a (finite) lattice is a pre-confluence with a minimum. The structure has a partial join operator:

Proposition 5 *For any $m \in \min(F)$ and any $x, y \in F^m$ their least upper bound is the least element of $F^x \cap F^y$ we further denote by $x \vee_F y$.*

This means that a pre-confluence is a union of lattices in which joins coincide. A particular case is which of a pre-confluence included in a host lattice and which is join preserving:

Definition 6 Let T be a lattice and $F \subseteq T$ be a pre-confluence with as join $\vee_F = \vee_T$, F is called a confluence of T.

An abstraction of T, as defined above is a confluence of T with $\perp_T$ as minimum. We have then the following property when considering 2^O as the host lattice:

Proposition 6 *Let* $X = 2^O$ *be a lattice,* $F \subseteq X$ *is a confluence of* X *if and only if for any* $x, y \in F^m$ *with* $m \in \min(F)$, *we have that* $x \cup y$ *belongs to* F.

A confluence is then associated with a set of interior operators:

Proposition 7 *Let* F *be a confluence of a lattice* X, *and* $m \in \min(F)$,

- $p_m : X^m \to X^m$, *such that* $p_m(x) = \vee_{q \in F^m \cap X_x} q$, *is an interior operator and* $p_m[X^m] = F^m$.

We are concerned here with extensional confluences, i.e., confluences of $X = 2^O$ [21] that generalize extensional abstractions as graph abstractions. In this case, let x be an element of X greater than or equal to some minimal element m of F, then $p_m(x)$ returns the greatest subset of x in F that includes m. We now define graph confluences, which are the original motivation for defining confluences:

Definition 7 Let $G = (O, E)$ be a graph, and F be the family of vertex subsets inducing connected subgraphs of G. F is a confluence of 2^O called the graph confluence of G.

Proof By definition, any singleton $\{s\}$ induces a connected subgraph of G. Furthermore, the union of two connected vertex subsets that each includes a given vertex singleton $\{s\}$ also is a connected vertex subset. Following Proposition 6, F is then a confluence of 2^O.

The elements of F are simply called the *connected vertex subsets* of O. By abuse of notation we write p_s and F^s rather than $p_{\{s\}}$ and $F^{\{s\}}$. The interior operator p_s projects then any vertex subset e containing vertex s on the connected component of the subgraph G_e induced by e that contains s. The up set F^s is then the set of connected vertex subsets containing s, and the union of all these F^s represents the whole set of connected subgraphs of G.

Example 3 Let $G = (O, E)$ be a graph (displayed at the bottom of Fig. 7) whose vertex set is $O = \{1, 2, 3, 4\}$. Let $F \subseteq 2^O$ be the set of connected vertex subsets of G. F is a confluence whose set of minimal elements is $\min(F) = \{\{1\}, \{2\}, \{3\}, \{4\}\}$. The subset $F^{1+3} = F^1 \cup F^3$ representing connected vertex subsets containing vertices 1 or 3 is also a confluence. Figure 7 displays the diagram of F^{1+3}. □

The extension e of a pattern q may then be projected through interior operators on various smaller *local extensions* $\{e_i\}$ corresponding to the connected components of the pattern subgraph. These interior operators are associated with *local closure operators* [21]:

Proposition 8 *Let* F *be a confluence of* $X = 2^O$, *m a minimal element of* F, *and* $L_{int(m)}$ *be the down set of the pattern lattice* L *whose elements* q *are such that* $q \geq int(m)$, *then*

$f_m = \text{int} \circ p_m \circ \text{ext}$ *is a closure operator on* $L_{int(m)}$

Fig. 7 A square graph (in the *bottom* of the figure) and the Hasse diagram of the confluence F^{1+3} of connected vertex subsets that contain vertices 1 or 3

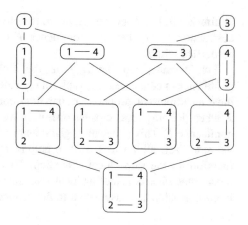

In a graph confluence, let $e = \text{ext}(q)$, then $p_s(e)$ is the connected component of the pattern subgraph G_e to which the vertex s belongs. Obviously, $p_s(e) = p_v(e)$ for any vertex v in the same connected component. Therefore $f_s(q)$ is a *local closed pattern* w.r.t. any vertex in this connected component, i.e., the most specific pattern shared by the vertices in the connected component.

Now a general result is that the set of local extensions is a pre-confluence:

Theorem 1 *The mapping* $h : F \to F : h(e) = p_m \circ \text{ext} \circ \text{int}(e)$ *for* $m \leq e$ *is a closure operator on F and* $E = h[F]$ *is a pre-confluence.*

$h(e)$ is therefore the local extension of $\text{int}(e)$ that contains $m \leq e$ and $h[F]$ is a pre-confluence isomorphic to the set P of *local concept pairs* defined as follows:

Definition 8 The set of local concept pairs $P = \{(e, l) \mid e = p_m \circ \text{ext}(l), l = \text{int}(e), m \leq e\}$ is called a *local concept pre-confluence*.

To summarize we have defined local concepts as (local extension, local closed patterns) pairs and we have shown that they are organized in a structure with possibly several minimal elements, therefore generalizing the concept lattice definition. In the graph confluence exemplified above the local extensions simply are the connected components of the pattern subgraphs. We will now extend graph confluences by intersecting graph confluences with abstractions.

4.1 Cc-Confluences

We remark now that we can freely intersect confluences:

Proposition 9 *Let* F_1 *and* F_2 *be confluences of X, then* $F_1 \cap F_2$ *is a confluence.*

Since abstractions of X are confluences of X with the bottom element of X as their unique minimal element, the above proposition means we can freely intersect abstractions with confluences to build smaller confluences. Many confluences can then be derived from a graph confluence by intersecting it with some abstractions. We call this family of confluences the *cc-confluences*.

Definition 9 Let F be the graph confluence of some graph G and A be a graph abstraction of G, then the confluence $F \cap A$ is called the cc-confluence of G associated with A.

For instance, considering A as the k-clique abstraction, we obtain the cc-confluence of connected subgraphs of G made of k-cliques. Note that cc-confluences have an important property: rather than considering the minimal elements m of F when defining local closure operators we can consider vertices as in the graph confluence F. This is because, given any abstract subgraph and any m included in its vertex set, all vertices v of m belong to the same connected component and therefore to the same local extension. This is computationally important as this means that when considering local extensions we only need to consider each vertex in the extension and associate it to the connected component to which it belongs.

4.2 Local Implications

Inclusion of local extensions defines validity of local implications $\square_m^A q \rightarrow \square_m^A w$, where m is a minimal element of F_A in the extension of q. Note that, as the local extension of pattern q is obtained by applying an interior operator, which is monotone, to the support set of q, we have that, whenever $\square^A q \rightarrow \square^A w$ is valid and $m \subseteq ext_A(w)$, we also have that $\square_m^A q \rightarrow \square_m^A w$ is valid, i.e., we may infer the latter local implication from the former abstract implication.

Example 4 Consider the graph displayed in Fig. 8. The 3-clique cc-confluence has as minimal elements $\{123, 567, 678\}$ and rewrites as $F_A = \{123, 567, 678, 5678\}$. The extension of pattern b is equal to its abstract extension $123, 678$, and the abstract closed pattern is also b. However, the corresponding abstract subgraph displays two connected components 123 and 678. The vertices of the latter share bc which is consequently its local closed pattern. This leads to a local implication:

- $\square_{678}^A b \rightarrow \square_{678}^A bc$

In a cc-confluence the local implication may be indexed with respect to any vertex of the corresponding connected component: a triangle in the same connected

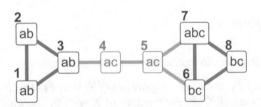

Fig. 8 The pattern b 3-clique abstract subgraph displays two connected components. The *blue one*, on the *left*, is also the pattern b abstract subgraph of motif a leading to the local implication $\square_{123}^A a \rightarrow \square_{123}^A ab$

component as 6, when considering the pattern b abstract subgraph, also has c, which rewrites as $\Box_6^A b \to \Box_6^A bc$. We will simply say that "a triangle containing 6 and which has b also has c."

We search now for a basis B_{F_A} of valid local implications from which we may infer any local implications. We will consider a basis B_m for a given minimal element m of F and obtain the whole basis $B_{F_A} = \cup B_m$ by joining these bases. Consider a given abstract closed pattern c whose abstract extension has a connected component that contains m, and let $l = f_m(c)$ be the corresponding local closed pattern, with respect to the cc-confluence F_A. This means that the implication $\Box_m c \to \Box_m l$ holds. We select then a basis B_m of *informative* ($l \neq c$) and *irredundant* (there is no other implication $\Box_m c' \to \Box_m l$ with c' less specific than c in the implication set) ones. From B_m we may infer all local implications associated with m by applying standard axioms in the same way as in the case of the *min–max basis* in the standard closed or abstract framework. The basis $B_{F_A} = \cup B_m$ represents the local knowledge deriving from the reduction of the extensional space from abstraction A to cc-confluence F_A:

Definition 10 The Local Min–max Basis B_{F_A} associated with the cc-confluence F_A is defined as:
$B_{F_A} = \{\Box_m^A c \to \Box_m^A l \mid$ where c A-closed, l locally closed, $c \neq l, \mathrm{ext}_m^A(c) = \mathrm{ext}_m^A(l),$
and for all $c' \subset c$ we have $\mathrm{ext}_m^A(c') \neq \mathrm{ext}_m^A(c)\}$

4.3 Interestingness Measures on Local Patterns and Implications

As in the abstract case, we may measure novelty brought locally [25]. We first extend the specificity measure to local patterns. If the ratio of a local extension to the abstract extension is high this means that the corresponding connected component is the largest part of the abstract subgraph.

Definition 11 Let q be a pattern, F be a cc-confluence and $m \in F$ be such that $m \subseteq \mathrm{ext}_A(q)$, the specificity of q near m is defined by

$$S_F(q, m) = \frac{|\mathrm{ext}_m^A(q)|}{|\mathrm{ext}_A(q)|}$$

We then define the informativity of a local implication by observing that in a valid local implication the left and right parts have same local extensions, while their abstract extensions are different. Therefore, as in the abstract knowledge case, we define the local informativity as the probability, at the abstract level, not to have the right part when the left one is true:

Definition 12 The local informativity of valid local implication $r : \square_m^A q \to \square_m^A qw$ is defined by

$$I_F(r) = 1 - \frac{|\operatorname{ext}_A(qw)|}{|\operatorname{ext}_A(q)|}$$

When local informativity is close to 1, it means that the probability to observe w when we have q at the abstract level is low: to be able to deduce w from q is specific of the graph region where m lies.

Example 5 We carry on Example 4 displayed in Fig. 8. The local specificity near 678 of pattern bc is $S_F(bc, 678) = 3 \div 3 = 1$: it does not appear elsewhere in the bc abstract subgraph. Informativity of local implication $\square_{678}^A b \to \square_{678}^A bc$ is $1 - (2 \div 6) = 0, 5$: the abstract support of b is 6, but only three vertices share the local closed pattern bc. Regarding the local closed pattern ab, it also has local specificity 1 as ab is only found in the connected component 123, and it leads to a new local implication $\square_{123}^A b \to \square_{123}^A ab$ whose informativity is $0, 5$. What we see here is that we obtain new knowledge regarding pattern b which depends on the region of the graph we consider.

Now, when considering the Teenage Friends attributed graph displayed in Fig. 3, clearly the friendship relations are organized in 3-cliques, therefore any stronger abstraction will be poorly informative. However, as mentioned in Sect. 1, when considering the 3-clique abstract graph associated with the empty pattern the unique connected component could be separated in several (overlapping) communities (displayed in Fig. 3 in various colors). We discuss and exemplify in the next section how to apply the local closure strategy to discover such sub-communities in an attributed graph.

5   Local Concepts from a Derived Graph

In what follows, we will consider a family $T \subseteq 2^O$ of vertex subsets, and consider T as the vertex set of a graph $G_T = (T, E_T)$ *derived* from G. The simple graph confluence F of 2^T is then the new extensional space and we will search for the corresponding local closed patterns. The local extensions are afterwards transformed into extensions in 2^O. Let $u : 2^T \to 2^O$ be such that $u(e_T) = \cup_{t \in e_T} t$. $u(e_T)$ is called the *flattening* of e_T. We consider then the two maps ext_T and int_T defined as follows:

- $\operatorname{ext}_T : L \to 2^T$ with $\operatorname{ext}_T(q) = \{t | t \subseteq \operatorname{ext}(q)\}$
- $\operatorname{int}_T : 2^T \to L$ with $\operatorname{int}_T(e_T) = \operatorname{int} \circ u(e_T)$

$\operatorname{ext}_T(q)$ represents the extension of q in T when considering that q occurs in t whenever q occurs in all elements of t (seen as a subset of O). Conversely $\operatorname{int}_T(e_T)$ represents the greatest pattern in L whose extension in T includes e_T, i.e., whose extension in O contains, as subsets, the elements of e_T. Now, consider as T the family of k-cliques of G and that $(t_1, t_2) \in E_T$ whenever t_1 and t_2 share $k-1$ vertices in G. A k-community in G [13] is a vertex subset that results from the flattening (in the sense defined above) of some connected component of G_T. The local closed patterns w.r.t. F are then most specific patterns occurring in k-communities of pattern subgraphs of G. This way we obtain local concepts and associated local implications, whose local extensions are these k-communities. Note that, in the derived case, the local concepts do not form a pre-confluence: technically we obtain a pre-confluence of 2^T, but two different local extensions in 2^T may result in the same flattening, corresponding to one 3-community. As a consequence, the local concept order is no more a pre-confluence.

5.1 Experiments on a Derived Graph

Coming back to our Teenage Friendship attributed graph, we have applied this strategy and built the derived graph G_T, where T is the set of 3-cliques of the original attributed graph. In Figs. 9 and 10 we display the ordered set of 3-communities with

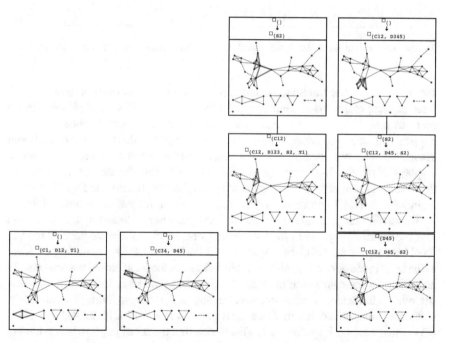

Fig. 9 The ordered set of size ≥ 4 3-communities of the Teenage Friendship network (part-I). The 3-communities are displayed in *red* and *bold lines* from the larger ones on the *top* to the smaller one on the *bottom*. The abstract subgraphs are displayed in *plain lines*. On the *right* at the *bottom* we have a 3-community displayed twice as it is built from two different abstract closed patterns

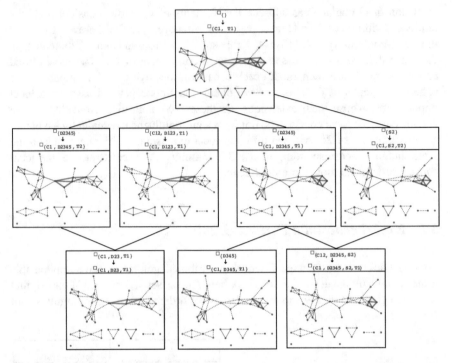

Fig. 10 The ordered set of size ≥ 4 3-communities of the Teenage Friendship network (part-II)

size at least 4.[7] The minimal 3-communities are the lowest ones on both figures. Each element of the pre-confluence represents a (3-community, local closed pattern) pair but may be associated with several non-redundant local implications. This happens for one 3-community displayed on the right at the bottom of Fig. 9 and associated with two local implications each represented in a square. Each square displays in red the 3-community, and in red+green+blue the abstract extension of the abstract closed pattern forming the left part of the implication. In Fig. 10 we have a unique maximal 3-community on the top, and a hierarchy of sub-communities.

We now investigate interestingness of local patterns and local implications. Note that the definitions given in Sect. 4.3 have to be adapted since we have to replace local extensions as defined in Sect. 4 by 3-communities, i.e., flattening of local extensions in the derived graph G_T. Table 4 displays the local closed patterns ranked according to their specificities. Each local closed pattern is indexed by the first triangle, in the lexicographic ordering, leading to the corresponding 3-community. Consider, for instance, the first two local patterns l_1 = D45-C12-S2 and l_2 = D45-C34 which have both specificity 1. This means that the set of pupils triangle having

---

[7]Formally, this means that we also apply an abstraction to the derived graph to avoid connected components corresponding to 3-communities smaller than four members.

Table 4 Top 15 local closed patterns ranked according to their specificity in the Teenage Friendship network

N°	$\square_m^A l$	$|ext_m^A(l)|$	$|ext^A(l)|$	$S_F(l,m)$
1	$\square_{25,31,32}^A$D2345-C1-S2-T1	5	5	1
2	$\square_{21,31,32}^A$C1-S2-T1	6	6	1
3	$\square_{17,19,24}^A$D23-C1-T1	4	4	1
4	$\square_{27,29,30}^A$D123-C12-S2-T1	4	4	1
5	$\square_{22,25,31}^A$D345-C1-T1	4	4	1
6	$\square_{10,11,15}^A$D45-C12-S2	4	4	1
7	$\square_{26,44,7}^A$D45-C34	5	5	1
9	$8\square_{40,45,46}^A$D12-C1-T1	4	6	0.667
10	$\square_{17,19,24}^A$C1-T1	11	17	0.647
11	$\square_{22,25,31}^A$D2345-C1-T1	6	10	0.6
12	$\square_{1,11,14}^A$D345-C12	6	10	0.6
13	$\square_{17,18,19}^A$D2345-C1-T1	5	10	0.5
14	$\square_{46,48,49}^A$D12-C1-T1	3	6	0.5
15	$\square_{17,19,24}^A$D123-C1-T1	5	11	0.455

D45-C12-S2 (respectively, D45-C34) forms a (unique) 3-community we further refer to as *Community 1* (respectively, *Community 2*). Communities 1 and 2 have in common high alcohol consumption behavior (D45), but differ in that the members of Community 1 do not smoke cannabis (C12) and have a regular sporting activity (S2), while the members of Community 2 have regular cannabis consumption (C34).

Consider now the implications of the local min–max basis, ranked according to their informativities, displayed in Table 5. We have here some implications with high informativity. As an example, the fifth (I_5) and ninth (I_9) local implications concern Community 1 while the second (I_2) concerns Community 2.

As an illustration we consider Community 1 and its two related local implications I_5 and I_9. As associated with the same community their rightmost member is the same local closed pattern D45-C12-S2. However, as they are extracted from different abstract subgraphs corresponding, respectively, to abstract closed patterns S2 and D45, they have different informativities. I_5 has informativity 0.75, while I_9 has informativity $\approx$ 0.56. In Fig. 11 we display Community 1 together with the necessary information to compute the local informativity of I_5.

The high informativity value of I_5 means that what the pupils in this community have in common, i.e., high alcohol consumption, no cannabis consumption and regular sporting activity (D45-C12-S2) is unfrequent among pupils in triangles with regular sporting activity outside of the community.

Table 5 Top 15 local implications of the min–max basis ranked according to their informativity int the Teenage Friendship network

N°	$\square^A_m c \to \square^A_m 1$	$|ext^A(l)|$	$|ext^A(c)|$	$I_F(r)$
1	$\square^A_{27,29,30}$C12 → $\square^A_{27,29,30}$D123-C12-S2-T1	4	27	0.852
2	$\square^A_{26,44,7}\emptyset$ → $\square^A_{26,44,7}$D45-C34	5	32	0.844
3	$\square^A_{46,48,49}\emptyset$ → $\square^A_{46,48,49}$D12-C1-T1	6	32	0.813
4	$\square^A_{40,45,46}\emptyset$ → $\square^A_{40,45,46}$D12-C1-T1	6	32	0.813
5	$\square^A_{10,11,15}$S2 → $\square^A_{10,11,15}$D45-C12-S2	4	16	0.75
6	$\square^A_{22,25,31}$D345 → $\square^A_{22,25,31}$D345-C1-T1	4	15	0.733
7	$\square^A_{1,11,14}\emptyset$ → $\square^A_{1,11,14}$D345-C12	10	32	0.688
8	$\square^A_{21,31,32}$S2 → $\square^A_{21,31,32}$C1-S2-T1	6	16	0.625
9	$\square^A_{10,11,15}$D45 → $\square^A_{10,11,15}$D45-C12-S2	4	9	0.556
10	$\square^A_{22,25,31}$D2345 → $\square^A_{22,25,31}$D2345-C1-T1	10	21	0.524
11	$\square^A_{17,18,19}$D2345 → $\square^A_{17,18,19}$D2345-C1-T1	10	21	0.524
12	$\square^A_{11,19,30}\emptyset$ → $\square^A_{11,19,30}$S2	16	32	0.5
13	$\square^A_{17,19,24}\emptyset$ → $\square^A_{17,19,24}$C1-T1	17	32	0.469
14	$\square^A_{11,19,30}$C12 → $\square^A_{11,19,30}$C12-S2	15	27	0.444
15	$\square^A_{25,31,32}$D2345-C12-S2 → $\square^A_{25,31,32}$D2345-C1-S2-T1	5	9	0.444

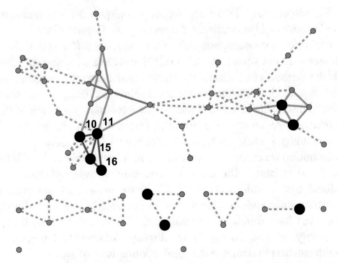

Fig. 11 Community 1 (represented as *bold dots* joined by *bold lines*) composed of five pupils with high alcohol consumption, no cannabis consumption C12 and regular sporting activity (D45-C12-S2). The community has size 4 and is one of the communities of the S2 abstract subgraph which is represented with *gray plain lines* joining 16 vertices. The Informativity of the local implication I_5 associated with community 1, which has S2 as its leftmost member and D45-C12-S2 as its rightmost member, is therefore $1 - 4/16 = 0.75$. Note also the *three black dots* representing vertices which have the S2 pattern but do not belong to its abstract subgraph

6 Implementation

In our experiments we first used the CORON software [28] to compute frequent closed patterns, according to some frequency threshold, then apply a set of PYTHON functions as a post-processing[8] to compute abstract and local patterns and implications. More recently we have implemented an efficient algorithm using a divide and conquer strategy similar to that proposed in [5] and implemented in [12]. This allows in particular to directly apply the frequency constraints at the abstract and local levels. A first version, named ParaminerLC, is experimented in [26] and was designed to handle the 3-communities local knowledge extraction problem. The selection of the implications belonging to the local min–max local basis is performed as a post-processing. Our current implementation in progress is a versatile program enumerating abstract and local frequent closed patterns and local implications with various definition of abstraction and locality.

7 Conclusion

In this article we have addressed problems in which the extensional space, made of the vertex subsets of an attributed network, is constrained according to connectivity properties. We have first considered abstract vertex subsets in which a constraint has to be satisfied by each vertex in the subgraph they induce, as, for instance, a minimum degree constraint. The extensional space is in this case a particular lattice called an abstraction. We have then shown, benefiting from previous work in FCA, how abstract support closed patterns, i.e., maximal patterns among those sharing the same abstract extension, could be obtained using a closure operator. This has resulted in defining a wide class of abstract concept lattices, whose elements are (abstract extension, abstract closed pattern) pairs, each corresponding to a particular abstraction. This way we obtain a global information on how the graph topology is related to the pattern extensions. We have then considered a way to extract local knowledge from an attributed network. For that purpose, using a recent extension of FCA to local extensional spaces, called confluences, we have related each pattern to various local extensions, corresponding to connected components in subgraphs induced by abstract vertex subsets. We obtain this way a set of local concepts, organized in a generalization of the lattice structure called a pre-confluence. Furthermore we have defined both abstract implications and local implications representing knowledge which is valid at the abstract and local levels, i.e., regarding the latter, in the vicinity of particular vertices. For both abstract and local patterns and implications we have proposed proper interestingness measures, namely specificity which measures to what extent the original extensions of patterns

---

[8]The corresponding software is to be found in https://lipn.univ-paris13.fr/~santini/data/ProjClos. tgz.

are preserved in abstract and local concepts, and informativity, which measures novelty brought by abstract and local implications. Finally we have applied these ideas to enumerate 3-communities in a network. These 3-communities are in fact sub-communities as each is a 3-community in some subnetwork induced by an attribute pattern.

Overall, what we propose here is a new way, brought by recent developments in Formal Concept Analysis, to explore social and complex networks as attributed graphs. As an application, we are currently involved in the ADALAB project which aims at helping the robot scientist EVE [30] to design experiments.[9] In this context, we use our methodology to explore a co-regulation network labeled with information regarding gene expression. Future works concerns, on the extensional side, applying these ideas to attributed directed graphs or multiplex networks. We also consider to use abstract and local extensional constraints while extending the pattern language to a wider class of pattern languages. First, as in [7, 9, 15] by building a meet-semilattice adapted to the mining problem and using interior operators to reduce it to a tractable language. This has been in particular successfully applied to graph mining [10]. Then, as in [5, 20] by considering confluent languages allowing to treat connectivity within the pattern language.

Acknowledgements This work was partially supported by CHIST-ERA grant (AdaLab, ANR 14-CHR2-0001-01).

References

1. Balasundaram, B., Butenko, S., Trukhanov, S.: Novel approaches for analyzing biological networks. J. Comb. Optim. **10**, 23–39 (2005)
2. Barabàsi, A.L., Albert, R.: Emergence of scaling in random networks. Science **286**(5439), 509–512 (1999). doi:10.1126/science.286.5439.509. http://www.sciencemag.org/content/286/5439/509.abstract
3. Batagelj, V., Zaversnik, M.: Fast algorithms for determining (generalized) core groups in social networks. Adv. Data Anal. Classif. **5**(2), 129–145 (2011). doi:10.1007/s11634-010-0079-y
4. Bechara Prado, A., Plantevit, M., Robardet, C., Boulicaut, J.F.: Mining graph topological patterns: finding co-variations among vertex descriptors. IEEE Trans. Knowl. Data Eng. **25**(9), 2090–2104 (2013). http://liris.cnrs.fr/publis/?id=5685
5. Boley, M., Horváth, T., Poigné, A., Wrobel, S.: Listing closed sets of strongly accessible set systems with applications to data mining. Theor. Comput. Sci. **411**(3), 691–700 (2010)
6. Borgatti, S.P., Everett, M.G.: Models of core/periphery structures. Soc. Netw. **21**(4), 375–395 (2000). doi:10.1016/S0378-8733(99)00019-2. http://dx.doi.org/10.1016/S0378-8733(99)00019-2
7. Ferré, S., Ridoux, O.: An introduction to logical information systems. Inf. Process. Manag. **40**(3), 383–419 (2004)
8. Fortunato, S.: Community detection in graphs. Phys. Rep. **486**(3–5), 75 – 174 (2010)

---

[9]http://www.adalab.mib.manchester.ac.uk/.

9. Ganter, B., Kuznetsov, S.O.: Pattern structures and their projections. In: International Conference on Conceptual Structures (ICCS). LNCS, vol. 2120, pp. 129–142. Springer, Heidelberg (2001)

10. Kuznetsov, S.O., Samokhin, M.V.: Learning closed sets of labeled graphs for chemical applications. In: Kramer, S., Pfahringer, B. (eds.) ILP. Lecture Notes in Computer Science, vol. 3625, pp. 190–208. Springer, Heidelberg (2005)

11. Mougel, P.N., Rigotti, C., Gandrillon, O.: Finding collections of k-clique percolated components in attributed graphs. In: PAKDD(2), Advances in Knowledge Discovery and Data Mining - 16th Pacific-Asia Conference, PAKDD 2012, Kuala Lumpur, 29 May–1 June 2012. Lecture Notes in Computer Science, vol. 7302, pp. 181–192. Springer, Heidelberg (2012)

12. Negrevergne, B., Termier, A., Rousset, M.C., Méhaut, J.F.: Paraminer: a generic pattern mining algorithm for multi-core architectures. Data Min. Knowl. Discov. **28**(3), 593–633 (2013). doi:10.1007/s10618-013-0313-2

13. Palla, G., Derenyi, I., Farkas, I., Vicsek, T.: Uncovering the overlapping community structure of complex networks in nature and society. Nature **435**(7043), 814–818 (2005). doi:10.1038/nature03607

14. Pasquier, N., Taouil, R., Bastide, Y., Stumme, G., Lakhal, L.: Generating a condensed representation for association rules. J. Intell. Inf. Syst. **24**(1), 29–60 (2005)

15. Pernelle, N., Rousset, M.C., Soldano, H., Ventos, V.: Zoom: a nested Galois lattices-based system for conceptual clustering. J. Exp. Theor. Artif. Intell. **2/3**(14), 157–187 (2002)

16. Rombach, M.P., Porter, M.A., Fowler, J.H., Mucha, P.J.: Core-periphery structure in networks. SIAM J. Appl. Math. **74**(1), 167–190 (2014). doi:10.1137/120881683. http://dx.doi.org/10.1137/120881683

17. Seidman, S.B.: Network structure and minimum degree. Soc. Netw. **5**, 269–287 (1983)

18. Silva, A., Meira Jr., W., Zaki, M.J.: Mining attribute-structure correlated patterns in large attributed graphs. Proc. VLDB Endow. **5**(5), 466–477 (2012). http://dl.acm.org/citation.cfm?id=2140436.2140443

19. Silva, A., Meira Jr., W., Zaki, M.J.: Mining attribute-structure correlated patterns in large attributed graphs. Proc. VLDB Endow. **5**(5), 466–477 (2012)

20. Soldano, H.: Closed patterns and abstraction beyond lattices. In: Glodeanu, C.V., Kaytoue, M., Sacarea, C. (eds.) Formal Concept Analysis 12th International Conference, ICFCA 2014, Cluj-Napoca, 10–13 June 2014. Lecture Notes in Computer Science, vol. 8478, pp. 203–218. Springer, Heidelberg (2014). doi:10.1007/978-3-319-07248-7_15. http://dx.doi.org/10.1007/978-3-319-07248-7_15

21. Soldano, H.: Extensional confluences and local closure operators. In: Baixeries, J., Sacarea, C., Ojeda-Aciego, M. (eds.) Formal Concept Analysis - Proceedings of the 13th International Conference, ICFCA 2015, Nerja, 23–26 June 2015. Lecture Notes in Computer Science, vol. 9113, pp. 128–144. Springer, Heidelberg (2015)

22. Soldano, H.: Extensional confluences and local closure operators. In: Baixeries, J., Sacarea, C., Ojeda-Aciego, M. (eds.) Formal Concept Analysis 13th International Conference, ICFCA, Nerja. LNCS, vol. 9113, pp. 128–144. Springer, Heidelberg (2015)

23. Soldano, H., Santini, G.: Graph abstraction for closed pattern mining in attributed network. In: Schaub, T., Friedrich, G., O'Sullivan, B. (eds.) European Conference in Artificial Intelligence (ECAI). Frontiers in Artificial Intelligence and Applications, vol. 263, pp. 849–854. IOS Press, Prague (2014)

24. Soldano, H., Ventos, V.: Abstract Concept Lattices. In: Valtchev, P., Jäschke, R. (eds.) International Conference on Formal Concept Analysis (ICFCA). LNAI, vol. 6628, pp. 235–250. Springer, Heidelberg (2011)

25. Soldano, H., Santini, G., Bouthinon, D.: Abstract and local rule learning in attributed networks. In: Esposito, F., Hacid, M.S., Pivert, O., Ras, Z. (eds.) Foundations of Intelligent Systems 22nd International Symposium, ISMIS, Lyon. LNAI, vol. 9384, pp. 313–323. Springer, Heidelberg (2015)

26. Soldano, H., Santini, G., Bouthinon, D.: Local knowledge discovery in attributed graphs. In: A. Esposito (ed.) 27th IEEE International Conference on Tools with Artificial Intelligence, ICTAI, pp. 250–257. IEEE Computer Society, Vietri sul Mare (2015)
27. Stumme, G., Taouil, R., Bastide, Y., Pasquier, N., Lakhal, L.: Computing iceberg concept lattices with titanic. Data Knowl. Eng. **42**(2), 189–222 (2002)
28. Szathmary, L., Napoli, A.: Coron: A framework for levelwise itemset mining algorithms. In: Ganter, B., Godin, R., Nguifo, E.M. (eds.) Third International Conference on Formal Concept Analysis (ICFCA'05), pp. 110–113, Lens, Supplementary Proceedings (2005)
29. Watts, D.J., Strogatz, S.H.: Collective dynamics of /'small-world/' networks. Nature **393**(6684), 440–442 (1998). http://dx.doi.org/10.1038/30918
30. Williams, K., Bilsland, E., Sparkes, A., Aubrey, W., Young, M., Soldatova, L.N., De Grave, K., Ramon, J., de Clare, M., Sirawaraporn, W., Oliver, S.G., King, R.D.: Cheaper faster drug development validated by the repositioning of drugs against neglected tropical diseases. J. R. Soc. Interface **12**(104) (2015). doi:10.1098/rsif.2014.1289. http://rsif.royalsocietypublishing.org/content/12/104/20141289

A Formal Concept Analysis Look at the Analysis of Affiliation Networks

Francisco J. Valverde-Albacete and Carmen Peláez-Moreno

1 Motivation and Introduction

Consider a *weighted bipartite graph*,[1] or *bipartite network* or *two-mode network*, $(G \cup M, R)$, where $R \in K^{G \times M}$ is a relation with values in an algebra K. This is a pervasive abstraction in Graph Theory [1] and Social Network Analysis (SNA) [2] where they are also known as *affiliation or membership networks*.

The *direct or dual-projection approach* is one of two competing methodologies for the analysis of two-mode networks, the other being the *conversion approach* [2]. In the latter, the data are first projected into a one-mode network and analysed with the tools of (weighted) graph analysis, that is (standard) network analysis. This raises evident and justified concerns of information loss [3].

In the dual-conversion approach, however, the analysis problem is transformed into two one-mode projection networks and analysed separately, with the projections on the rows P_G and the columns P_M being the matrices:

$$P_G = R \otimes R^{\mathrm{T}} \qquad\qquad P_M = R^{\mathrm{T}} \otimes R. \qquad (1)$$

The dual-projection approach postulates that we can provide measures of centrality, core vs. periphery and structural equivalence for each of the projection networks with limited loss of global information, in terms of the Singular Value Decomposition (SVD) [4]. This is a set of results about the decomposition of real- or

---

[1] We will consider all graphs in this paper as directed graphs unless otherwise stated.

F.J. Valverde-Albacete (✉) • C. Peláez-Moreno
Department of Signal Theory and Communications, Universidad Carlos III de Madrid, 28911 Leganés, Spain
e-mail: fva@tsc.uc3m.es; carmen@tsc.uc3m.es

© Springer International Publishing AG 2017
R. Missaoui et al. (eds.), *Formal Concept Analysis of Social Networks*,
Lecture Notes in Social Networks, DOI 10.1007/978-3-319-64167-6_7

complex-valued rectangular matrices [5, 6] with applications in data processing, signal theory, machine learning and computer science at large [7], the most important of which is the following:

Theorem 1 (The Singular Value Decomposition Theorem) *Given a matrix $M \in \mathcal{M}_{m \times n}(\mathcal{K})$ where $\mathcal{K}$ is a field, there is a factorization $M = U \Sigma V^*$—where $\cdot^*$ stands for conjugation—given in terms of three matrices*

- *$U \in \mathcal{M}_{m \times m}(\mathcal{K})$ is a unitary matrix of* left *singular vectors.*
- *$\Sigma \in \mathcal{M}_{m \times n}(\mathcal{K})$ is a diagonal matrix of non-negative real values called the singular values.*
- *$V \in \mathcal{M}_{n \times n}(\mathcal{K})$ is a unitary matrix of* right *singular vectors.*

Often the singular values of a matrix are listed in descending order—and the left and right singular vectors are re-ordered accordingly—as a prelude to any of a number of *reconstruction theorems* aimed at re-building the original matrix M from the triples of singular value, left and right vectors (σ_i, u_i, v_i) [see 5, 6, for details]. This is particularly interesting for model building.

1.1 The Analysis of Bipartite Networks with Formal Concept Analysis

Formal Concept Analysis (FCA) can be conceived as a data-driven unsupervised learning technique for Boolean data. Its main results can be summarized as follows [8].

Theorem 2 (Basic Theorem of FCA, Extended[2]) *Let G be a set of objetcs, M a set of attributes and (G, M, I) be a formal context with $I \subseteq G \times M$ and polar operators $\cdot^{\uparrow} : 2^G \to 2^M$ and $\cdot^{\downarrow} : 2^M \to 2^G$.*

$$A^{\uparrow} = \{m \in M \mid \forall g \in A, gIm\} \qquad B^{\downarrow} = \{g \in G \mid \forall m \in B, gIm\}$$

and call formal concepts the pairs with $A^{\uparrow} = B \Leftrightarrow A = B^{\downarrow}$ ordered as

$$(A_1, B_1) \leq (A_2, B_2) \Leftrightarrow A_1 \subseteq A_2 \Leftrightarrow B_1 \supseteq B_2.$$

Then:

1. *The set of formal concepts $\mathfrak{B}(G, M, I)$ with the hierarchical order is a complete lattice called the concept lattice of (G, M, I) in which infima and suprema are given by:*

---

[2]In [9] this is called "a fundamental pattern for the occurrence of lattices in general"; in [8] it is the "Basic Theorem of Concept Lattices", and in Chap. 3 of [10] it is "the fundamental theorem of concept lattices".

$$\bigwedge_{t \in T}(A_t, B_t) = \left(\bigcap_{t \in T} A_t, \left(\bigcup_{t \in T} B_t\right)^{\uparrow}\right) \quad \bigvee_{t \in T}(A_t, B_t) = \left(\left(\bigcup_{t \in T} A_t\right)^{\downarrow}, \bigcap_{t \in T} B_t\right)$$

(2)

2. Conversely, a complete lattice $\mathcal{V} = \langle V, \leq \rangle$ is isomorphic to $\mathfrak{B}(G, M, I)$ if and only if there are mappings $\overline{\gamma} : G \to V$ and $\overline{\mu} : M \to V$ such that $\overline{\gamma}(G)$ is supremum-dense in $\mathcal{V}$, $\overline{\mu}(M)$ is infimum-dense in $\mathcal{V}$ and gIm is equivalent to $\overline{\gamma}(g) \leq \overline{\mu}(m)$ for all $g \in G$ and all $m \in M$. In particular, $\mathcal{V} \cong \mathfrak{B}(V, V, \leq)$.

The ability to analyse bipartite digraphs (that is Boolean bipartite networks) comes from the existence of a "cryptomorphism" [11, pp. 155–156]—which we take to mean "unexpected isomorphisms"[3]—with formal contexts, possibly first identified in [13]. We adhere here to the advantages of using cryptomorphisms for the description of apparently different objects laid out in [14], namely *reaching a better understanding of the tools being analysed*.

The Cryptomorphism of Bipartite Digraphs and Formal Contexts When K is the Boolean set, consider the two following definitions:

- In Graph Theory, let G and M be two disjoint sets and consider the graph (V, E) with $V = G \cup M$ and $E \in 2^{V \times V}$ such that for every ordered pair $e = (g_e, m_e) \in E$ its endpoints belong to different subsets of V, $g_e \in G$ and $m_e \in M$. Then (V, E) is a *bipartite graph*.
- In FCA, let G be a set of objects and M be a set of attributes, and $I \in 2^{G \times M}$ be an *incidence relation* between them. Then (G, M, I) is a *formal context*.

Clearly, they define cryptomorphic entities whereby I is the restriction of E to $G \times M$ [15, Sect. 3.1].

Hence the capabilities of FCA for representing bipartite digraphs follow from the universal representation capabilities of concept lattices expressed in the Basic Theorem. This promises that many fundamental abstractions in each domain will have an important role in the other. Crucially, formal concepts are cryptomorphic to bicliques maximal with respect to inclusion [15, Proposition 1]. This was probably first suggested in [16], clearly stated in [13], an later taken up by a number of researchers [17, 18]. Note that these techniques and concerns pre-date the apparition of multi-valued extensions to FCA and therefore concentrate in Boolean data.

Some of the advantages of using FCA for modelling bipartite networks stem from the hierarchical, non-partitional (overlapping) clustering of both domains G and M [19]. This approach is contextualized and summarized in the wider context

---

[3]Birkhoff actually coined "crypto-isomorphism", but the term seems to have been forced to evolve [12]. We point out that the "surprise" must come from finding concepts of different subfields to be the same. Of course cryptomorphisms boil down to plain isomorphisms as soon as the surprise fades away, so it is a mathematical concept more of an educational or sociological than a formal nature.

of Social Network Visualization in [20], whose most harrying problem is the visual clutter. For concept lattices this is mostly addressed by means of algebraic information reduction techniques [21] or heuristic pruning [19].

For our present purposes [22] has already noted that formal concepts define optimal factors for the reconstruction of Boolean matrices—whether incidences or adjacencies—and this is a result that strongly hints that FCA is related to the SVD.

SVD Leads to Non-Boolean Contexts The previous argumentation would seem to imply that the SVD, as a technique to analyse weighted bipartite digraphs, is also important for the analysis of formal contexts and concept lattices. Alas, although the SVD is cursorily applied to bipartite digraphs whose entries belong to $\{0, 1\}$, it is actually a procedure developed for matrices with entries in the complex field $\mathbb{C}$. To draw a parallel with the nice situation in the previous section would call for the consideration of an isomorphism between bipartite digraphs whose edge weights belong to an algebra and *multi-valued formal contexts* with incidences in said algebra.

Unfortunately, the theory of concept lattices issuing from such multi-valued contexts is not as complete as (standard) FCA: for instance, when the context takes values in a fuzzy-semiring the universal representation capabilities in the Main Theorem [23, Theorem 5.3] have not been cast in terms of (fuzzy) bipartite networks, to the best of our knowledge, and the Main Theorem of $\mathcal{K}$-FCA [24, Theorem 2.14], where $\mathcal{K}$ is a complete idempotent semifield (see below) has as yet unexplored representation capabilities. We note that, despite these limitations, a number of results on the reconstruction of matrices specifically based on fuzzy-formal concepts are available [25].

1.2 The Study of Networks Using HITS and the SVD

In this chapter we are interested in laying out the relationship between Formal Concept Analysis [8] and the dual-projection approach to the analysis of bipartite networks. At the beginning of this section we have argued that the SVD must figure prominently in this picture, so we will detour slightly to show yet one more instance of the pervasiveness of it in the analysis of networks: its relation to one of the first well-known approaches to link analysis on the Web, the HITS algorithm.

The Hubs and Authorities algorithm or Hyperlink-Induced Topic Search (HITS) [26, 27] was designed to solve the problem of ranking the nodes of a dynamic, directed 1-mode network of nodes obtained from a query against a search engine. It postulates the existence of two qualities in nodes: their *authoritativeness*—their quality of being authorities with respect to a pervasive topic in the nodes—and their *hubness*—their quality of being good pointers to authorities. These are now cursorily available in software for analysing network data, e.g. [28].

Consider a network, that is, a weighted directed graph, $N = (V, E, w)$ where $V = \{v_i\}_{i=1}^n$ is a set of nodes $E \subseteq V \times V$ is a set of edges, and $w : E \to [0, 1]$ a weight function on the edges $w(v_i, v_j) > 0$, $(v_i, v_j) \in E$ and $w(v_i, v_j) = 0$ otherwise. It can alternatively be defined by an adjacency matrix T_N with $[0, 1]$-weights for its edges, $(T_N)_{ij} = w(v_i, v_j)$. The notation tries to suggest that T_N is a *stochastic transition matrix* rather than a generic adjacency.[4]

HITS is solved in terms of the left and right eigenvectors of this matrix T_N, where the vertices and edges of the digraph model the nodes and links in any (social) network. HITS finds an authority score $a(v_i)$ and a hub score $h(v_i)$ for each node aggregated as vectors, based on the following iterative procedure:

- Start with initial vector estimates $h^{<0>}$ and $a^{<0>}$ of the hub and authority scores.
- Upgrade the scores with[5]:

$$h \leftarrow Ta \qquad\qquad a \leftarrow T^\mathsf{T} h \qquad (3)$$

so that in general, for $k \geq 1$:

$$h^{<k>} = (TT^\mathsf{T})^k h^{<0>} \qquad\qquad a^{<k>} = (T^\mathsf{T}T)^k a^{<0>}$$
$$h^{<k>} = (TT^\mathsf{T})^{k-1} Ta^{<0>} \qquad\qquad a^{<k>} = (T^\mathsf{T}T)^{k-1} T^\mathsf{T} h^{<0>} \qquad (4)$$

- Since matrix T is non-negative, in general the sequences $\{h^{<k>}\}_k$ and $\{a^{<k>}\}_k$ would diverge, so the next step is to prove that the limits:

$$\lim_{k \to \infty} \frac{h^{<k>}}{c^k} = h^{<*>} \qquad\qquad \lim_{k \to \infty} \frac{a^{<k>}}{d^k} = a^{<*>} \qquad (5)$$

exist, in which case they are eigenvectors of their respective matrices for seemingly arbitrary c and d,

$$(TT^\mathsf{T})h^{<*>} = ch^{<*>} \qquad\qquad (T^\mathsf{T}T)a^{<*>} = da^{<*>} . \qquad (6)$$

- As long as the initial estimates do not inhabit the null space of these matrices— making them orthogonal to $h^{<*>}$ and $a^{<*>}$, respectively—the iterative process will end up finding the principal eigenvectors. The proof of this fact entails that the initial estimates $h^{<0>}$ and $a^{<0>}$ should be non-negative.

It is easy to prove the following:

Lemma 1 *HITS is a specialized version of the SVD.*

---

[4]This procedure will be extended in Sect. 2.3.
[5]To lessen the visual clutter, we drop the graph index from the matrix.

Proof Since we want to emphasize the mutual dependence of hubness and authoritativeness scores, so after [29] we write (3) in matrix form

$$\begin{bmatrix} h \\ a \end{bmatrix} \leftarrow \begin{bmatrix} \cdot & T \\ T^\mathrm{T} & \cdot \end{bmatrix} \otimes \begin{bmatrix} h \\ a \end{bmatrix}$$

where we have substituted zero matrices for dots, as customary.

The arrows are used to suggest that we are interested in the fixpoint of the iterative update of this matrix equation. But we know that a fixpoint of it is the analytical solution of the following eigenproblem in the variable $z = [x^\mathrm{T} y^\mathrm{T}]^\mathrm{T}$,

$$A \otimes z = z \otimes \sigma \Leftrightarrow \begin{bmatrix} \cdot & T \\ T^\mathrm{T} & \cdot \end{bmatrix} \otimes \begin{bmatrix} x \\ y \end{bmatrix} = \begin{bmatrix} x \\ y \end{bmatrix} \otimes \sigma \tag{7}$$

for eigenvalue $\sigma = 1$.

To see that the solutions to this problem are of the form $w = [h^\mathrm{T} a^\mathrm{T}]^\mathrm{T}$ that is, pairs of hub and authority vectors, we expand the system (7) into two equations—called by Lanczos the "shifted eigenvalue problem" [29]—

$$T \otimes a = h \otimes \sigma \qquad\qquad T^\mathrm{T} \otimes h = a \otimes \sigma \tag{8}$$

Equation (8) is the proof that HITS is actually trying to solve the *singular value-singular vector problem* [5, 29], where h has the role of a left singular vector, a that of a right singular vector, and $\sigma = 1$ is the singular value. □

Note that:

- Under the conditions laid out in the original HITS setting, the singular value is not important.
- The projectors appear in the solution of (7) by pre-multiplying both sides of the equation with A and then we would obtain decoupled solutions that can be re-coupled with (8).

$$A^2 \otimes z = z \otimes \sigma^2 \Leftrightarrow \begin{bmatrix} TT^\mathrm{T} & \cdot \\ \cdot & T^\mathrm{T}T \end{bmatrix} \otimes \begin{bmatrix} x \\ y \end{bmatrix} = \begin{bmatrix} x \\ y \end{bmatrix} \otimes \sigma^2 \tag{9}$$

In light of this, we can see how *HITS, which in principle is available for 1-mode networks, that is, it is a* conversion approach *procedure is actually using a* direct or dual-projection approach *in considering both vector spaces associated with T.*

In light of this, we can see how *HITS is actually using a* direct or dual-projection approach—*in considering both vector spaces associated with T*—*in spite of being actually available for 1-mode networks, that is, being a* conversion approach.

1.3 The Problem and Reading Guide

In this paper we develop a similar tool as the SVD for bipartite networks with non-Boolean edge weights, but we develop it as if it were an instance of the better understood, HITS problem.

1. First, as suggested by the form of (4), we consider two sets G and M, because we want to study the quality of being a hub and being an authority separately. This implies passing from directed bipartite graphs to *networks,* also known as *weighted (directed) bipartite graphs.*
2. Second, the matrix algorithm HITS requires a "positive" algebra with addition, multiplication and scalar division—in the case of the original HITS, this semifield is $\mathbb{R}_+$, the positive reals. Hence we consider edge weights $R \in K^{G \times M}$ in a naturally ordered semiring with division or *positive semifield* $\mathcal{K}$ (cfr. Sect. 2.1). Then $\mathcal{K}$-formal contexts, denoted as (G, M, R), are a natural encoding for this type of weighted bipartite digraphs.

Note that the original HITS setting can be recovered by using $V := G = M$ and $T := R$ and working in the semifield of positive reals $\mathbb{R}_0^+$ with $R_{ij} = 1$ if $(v_i, v_j) \in E$ and $R_{ij} = 0$ otherwise. Similarly, the original dual-projection approach deals only with the case where R is actually binary but is considered to be embedded in the complex numbers.

To develop our program we first introduce in Sect. 2.1 some definitions and notation about semirings in general, and about positive semifields in particular. In Sect. 2.2 we introduce the eigenproblem over dioids as a step to solving the singular value problem in dioids, and in Sects. 2.3 and 2.4 a very general technique to do so. Section 3 presents the weight of our results, including a short Example and a Discussion. We finish with a Summary and Conclusions.

2 Theory and Methods

2.1 Semiring and Semimodules over Semirings

A *semiring* is an algebra $\mathscr{S} = \langle S, \oplus, \otimes, \epsilon, e \rangle$ whose additive structure, $\langle S, \oplus, \epsilon \rangle$, is a commutative monoid and whose multiplicative structure, $\langle S \setminus \{\epsilon\}, \otimes, e \rangle$, is a monoid with multiplication distributing over addition from right and left and an additive neutral element absorbing for $\otimes$, i.e. $\forall a \in S, \epsilon \otimes a = \epsilon$. A semiring is:

- *Commutative,* if its multiplication is commutative.
- *Zerosumfree,* if it does not have non-null additive factors of zero, $a \oplus b = \epsilon \Rightarrow a = \epsilon$ and $b = \epsilon$, $\forall a, b \in S$.
- *Entire,* if $a \otimes b = \epsilon \Rightarrow a = \epsilon$ or $b = \epsilon$, $\forall a, b \in S$.
- *Idempotent,* if its addition is.

- A *selective semiring,* if the arguments attaining the value of the additive operation can be identified.
- *Radicable,* if the equation $a^b = c$ can be solved for a.
- *Complete,* [30] if for every (possibly infinite) family of elements $\{a_i\}_{i \in I} \subseteq S$ we can define an element $\sum_{i \in I} a_i \in S$ such that

  1. if $I = \varnothing$, then $\sum_{i \in I} a_i = \epsilon$,
  2. if $I = \{1 \ldots n\}$, then $\sum_{i \in I} a_i = a_1 \oplus \ldots \oplus a_n$,
  3. if $b \in S$, then $b \otimes \left(\sum_{i \in I} a_i \right) = \sum_{i \in I} b \otimes a_i$ and $\left(\sum_{i \in I} a_i \right) \otimes b = \sum_{i \in I} a_i \otimes b$, and
  4. if $\{I_j\}_{j \in J}$ is a partition of I, then $\sum_{i \in I} a_i = \sum_{j \in J} \left(\sum_{i \in I_j} a_i \right)$.

Entire zerosumfree semirings are called sometimes *information algebras* and have abundant applications [31]. Their importance stems from the fact that they model *positive quantities.*

Crucially, every commutative semiring accepts a *canonical preorder,* as $a \leq b$ if and only if there exists $c \in S$ with $a \oplus c = b$ which is *compatible with addition.* A *dioid* is a commutative semiring where this relation is actually an order. Dioids are zerosumfree. A dioid that is also entire—that is, when $a \otimes b = \epsilon$ then either $a = \epsilon$ or $b = \epsilon$ or both—is a *positive dioid.*

If I is countable in the definitions above, then $\mathscr{S}$ is *countably complete* and already zerosumfree [32, Proposition 22.28]. The importance for us is that in complete semirings, the existence of the transitive closures is guaranteed (see Sect. 2.3). Commutative *complete* dioids are already complete residuated lattices.

A semiring whose commutative multiplicative structure is a group will be called a *semifield.*[6] Semifields are all entire, and we will use $\mathscr{K}$ to refer to them. Therefore semifield which is also a dioid is both entire and naturally ordered. These are sometimes called *positive semifields,* examples of which are the positive rationals, the positive reals or the max-plus and min-plus semifields. Semifields are all incomplete except for the Booleans, but they can be completed as $\overline{\mathscr{K}}$ [24], and we will not differentiate between *complete or completed* structures.

A *semimodule (over a semiring)* is an analogue of a vector space over a field. Semimodules inherit from their defining semirings the qualities of being zerosumfree, complete or having a natural order. In fact, semimodules over complete commutative dioids are also complete lattices. *Rectangular matrices over a semiring* form a semimodule $\mathscr{M}_{m \times n}(\mathscr{S})$, and in particular, row- and column-spaces $\mathscr{S}^{1 \times n}$ and $\mathscr{S}^{n \times 1}$. The set of square matrices $\mathscr{M}_n(\mathscr{S})$ is also a semiring (but non-commutative unless $n = 1$).

---

[6]This term is not standard: for instance, [33] prefer to use "semiring with a multiplicative group structure", but we prefer *semifield* to shorten out statements.

2.2 The Eigenvalue Problem over Dioids

Given (6), understanding the HITS iteration is easier once understood the eigenvalue problem in a semiring. So let $\mathcal{M}_n(\mathcal{S})$ be the semiring of square matrices over a semiring $\mathcal{S}$ with the usual operations. Given $A \in \mathcal{M}_n(\mathcal{S})$ the *right (left) eigenproblem* is the task of finding the *right eigenvectors* $v \in S^{n \times 1}$ and *right eigenvalues* $\rho \in S$ (respectively, *left eigenvectors* $u \in S^{1 \times n}$ and *left eigenvalues* $\lambda \in S$) satisfying:

$$u \otimes A = \lambda \otimes u \qquad\qquad A \otimes v = v \otimes \rho \qquad (10)$$

The left and right eigenspaces and spectra are the sets of these solutions:

$$\Lambda(A) = \{\lambda \in S \mid \mathcal{U}_\lambda(A) \neq \{\epsilon^n\}\} \qquad P(A) = \{\rho \in S \mid \mathcal{V}_\rho(A) \neq \{\epsilon^n\}\}$$

$$\mathcal{U}_\lambda(A) = \{u \in S^{1 \times n} \mid u \otimes A = \lambda \otimes u\} \quad \mathcal{V}_\rho(A) = \{v \in S^{n \times 1} \mid A \otimes v = v \otimes \rho\}$$

$$\mathcal{U}(A) = \bigcup_{\lambda \in \Lambda(A)} \mathcal{U}_\lambda(A) \qquad\qquad \mathcal{V}(A) = \bigcup_{\rho \in P(A)} \mathcal{V}_\rho(A) \qquad (11)$$

Since $\Lambda(A) = P(A^{\mathsf{T}})$ and $\mathcal{U}_A(A) = \mathcal{V}_\lambda(A^{\mathsf{T}})$, from now on we will omit references to left eigenvalues, eigenvectors and spectra, unless we want to emphasize differences.

In order to solve (7) in dioids we have to use the following theorem [33, 34]:

Theorem 3 (Gondran and Minoux [34, Theorem 1]) *Let* $A \in \mathcal{S}^{n \times n}$. *If* A^* *exists, the following two conditions are equivalent:*

1. $A_{\cdot i}^{+} \otimes \mu = A_{\cdot i}^{*} \otimes \mu$ *for some* $i \in \{1 \ldots n\}$, *and* $\mu \in S$.
2. $A_{\cdot i}^{+} \otimes \mu$ *(and* $A_{\cdot i}^{*} \otimes \mu$) *is an eigenvector of* A *for* e, $A_{\cdot i}^{+} \otimes \mu \in \mathcal{V}_e(A)$.

where we define the *transitive closure* $A^{+} = \sum_{k=1}^{\infty} A$ and the *transitive reflexive closure* $A^* = \sum_{k=0}^{\infty} A$ of A (also called *Kleene's plus and star operators*).

In [35–37] Gondran and Minoux' theorem was made more specific in two directions: on the one hand, by focusing on particular types of completed idempotent semirings—semirings with a natural order where infinite additions of elements exist, so transitive closures are guaranteed to exist and sets of generators can be found for the eigenspaces—and, on the other hand, by considering more easily visualizable subsemimodules than the whole eigenspace—a better choice for exploratory data analysis.

2.3 Graphs, Matrices and Closures over Dioids

From Theorem 3 it is clear that we need efficient methods to obtain the closures in order to solve the eigenvalue-eigenvector problem—and hence HITS and SVD— in the general setting of semirings. For this purpose, it is interesting to extend the cryptomorphism between weighted graphs and square matrices of Sect. 1.2 explicitly:

- For a matrix $A \in \mathcal{M}_n(\mathcal{S})$, the *network or weighted digraph induced by A*, $N_A = (V_A, E_A, w_A)$, consists of a set of vertices V_A, a set of arcs , $E_A = \{(i,j) \mid A_{ij} \neq \epsilon_S\}$, and a weight $w_A : V_A \times V_A \rightarrow S$, $(i,j) \mapsto w_A(i,j) = a_{ij}$.
- For a weighted directed graph, $N = (V, E, w)$ where $V = \{v_i\}_{i=1}^n$ is a set of nodes $E \subseteq V \times V$ is a set of edges, and $w : E \rightarrow S$ a weight function on the edges $w(v_i, v_j) > 0, (v_i, v_j) \in E$ and $w(v_i, v_j) = 0$ otherwise, the matrix $A_N \in \mathcal{M}_n(\mathcal{S})$ is defined as $(A_N)_{ij} = w(v_i, v_j)$.

This allows us *intuitively to apply all notions from networks to matrices and vice versa*, like the underlying graph $G_A = (V_A, E_A)$ disregarding the weights, the set of paths $\Pi_A^+(i,j)$ between nodes i and j or the set of cycles C_A^+. The following account is a summary of results in this respect, and we refer the reader to [35, 36] for proofs.

Lemma 2 *Let $A \in \mathcal{M}_n(S)$ be a square matrix over a commutative semiring $\mathcal{S}$. A^* exists if and only if A^+ exists and then:*

$$A^+ = A \otimes A^* = A^* \otimes A \qquad\qquad A^* = I \oplus A^+$$

But since in incomplete semirings the existence of the closures is not warranted, our natural environment should be that of *complete semirings*.

On the other hands, in dioids the following lemma holds:

Lemma 3 *Let $A \in \mathcal{M}_n(S)$ be a square matrix over a dioid $\mathcal{S}$. For partition $\bar{n} = \alpha \cup \beta$ call* PER $(A) = A_{\beta\alpha}A_{\alpha\alpha}^*A_{\alpha\beta} \oplus A_{\beta\beta}$. *Then*

$$\begin{pmatrix} A_{\alpha\alpha} & A_{\alpha\beta} \\ A_{\beta\alpha} & A_{\beta\beta} \end{pmatrix}^+ = \begin{pmatrix} A_{\alpha\alpha}^+ \oplus A_{\alpha\alpha}^*A_{\alpha\beta}\text{PER}\,(A)^*A_{\beta\alpha}A_{\alpha\alpha}^* & A_{\alpha\alpha}^*A_{\alpha\beta}\text{PER}\,(A)^* \\ \text{PER}\,(A)^*A_{\beta\alpha}A_{\alpha\alpha}^* & \text{PER}\,(A)^+ \end{pmatrix}$$

Proof Adapted from [38, Lemma 4.101]   □

Notice that closures and simultaneous row and column permutations commute:

Lemma 4 *Let $A, B \in \mathcal{M}_n(\mathcal{S})$ and let P be a permutation such that $B = P^{\mathsf{T}}AP$. Then $B^+ = P^{\mathsf{T}}A^+P$ and $B^* = P^{\mathsf{T}}A^*P$.*

A square matrix is *irreducible* if it cannot be simultaneously permuted into a triangular upper (or lower) form. Otherwise we say it is *reducible*. Irreducibility expresses itself as a graph property on the induced digraph D_A of Sect. 1.2.

Lemma 5 *If $A \in \mathcal{M}_n(S)$ is irreducible, then:*

- *The induced digraph D_A has a single strongly connected component.*
- *All nodes in its induced digraph D_A are connected by cycles.*

The irreducible case is used as a basic case in the recursive building of the closure of any possible matrix next. In it, the condensation digraph is built using the classes of the reachability relation in D_A as the vertices (the strongly connected components of D_A) and their connections as edges:

Lemma 6 (Recursive Upper Frobenius Normal Form, UFNF) *Let* $A \in \mathcal{M}_n(S)$ *be a matrix over a semiring and* $\overline{G}_A$ *its condensation digraph. Then,*

1. *(UFNF$_3$) If* A *has zero lines it can be transformed by a simultaneous row and column permutation of* V_A *into the following form:*

$$
P_3^{\mathrm{T}} \otimes A \otimes P_3 =
\begin{bmatrix}
\mathscr{E}_{\iota\iota} & \cdot & \cdot & \cdot \\
\cdot & \mathscr{E}_{\alpha\alpha} & A_{\alpha\beta} & A_{\alpha\omega} \\
\cdot & \cdot & A_{\beta\beta} & A_{\beta\omega} \\
\cdot & \cdot & \cdot & \mathscr{E}_{\omega\omega}
\end{bmatrix}
\tag{12}
$$

where either $A_{\alpha\beta}$ *or* $A_{\alpha\omega}$ *or both are non-zero, and either* $A_{\alpha\omega}$ *or* $A_{\beta\omega}$ *or both are non-zero. Furthermore,* P_3 *is obtained concatenating permutations for the indices of simultaneously zero columns and rows* V_ι, *the indices of zero columns but non-zero rows* V_α, *the indices of zero rows but non-zero columns* V_ω *and the rest* V_β *as* $P_3 = P(V_\iota)P(V_\alpha)P(V_\beta)P(V_\omega)$.

2. *(UFNF$_2$) If* A *has no zero lines it can be transformed by a simultaneous row and column permutation* $P_2 = P(A_1) \dots P(A_k)$ *into block diagonal UFNF:*

$$
P_2^{\mathrm{T}} \otimes A \otimes P_2 =
\begin{bmatrix}
A_1 & \cdot & \dots & \cdot \\
\cdot & A_2 & \dots & \cdot \\
\vdots & \vdots & \ddots & \vdots \\
\cdot & \cdot & \dots & A_K
\end{bmatrix}
\tag{13}
$$

where $\{A_k\}_{k=1}^K$, $K \geq 1$ *are the matrices of connected components of* $\overline{G}_A$.

3. *(UFNF$_1$) If* A *is reducible with no zero lines and a single connected component it can be simultaneously row- and column-permuted by* P_1 *to*

$$
P_1^{\mathrm{T}} \otimes A \otimes P_1 =
\begin{bmatrix}
A_{11} & A_{12} & \cdots & A_{1R} \\
\cdot & A_{22} & \cdots & A_{2R} \\
\vdots & \vdots & \ddots & \vdots \\
\cdot & \cdot & \cdots & A_{RR}
\end{bmatrix}
\tag{14}
$$

where A_{rr} *are the matrices associated with each of its* R *strongly connected components (sorted in a topological ordering), and* $P_1 = P(A_{11}) \dots P(A_{RR})$.

Note that irreducible blocks are the base case of UFNF$_1$, so we sometimes refer to irreducible matrices as being in UFNF$_0$.

Note that as a result of this Section, we know how to calculate algorithmically the transitive closure for any type of matrix A in any complete semiring.

2.4 Eigenvalues and Eigenvectors of Matrices over Complete Dioids

By the reasoning in previous sections and the cryptomorphism above, eigenvectors of the projection matrices in the dual-projection approach are vectors describing some qualities of the nodes in the weighted bipartite graph, e.g. authoritativeness or hubness in HITS. This is the reason why we need to characterize such vectors better.

2.4.1 Orthogonality of Eigenvectors

In spectral decomposition, orthogonality of the eigenvectors plays an important role. In zerosumfree semimodules orthogonality cannot be as prevalent as in standard vector spaces. To see this, first call the *support* of a vector, the set of indices of non-null coordinates $\mathrm{supp}(v) = \{i \in \bar{n} | v_i \neq \epsilon\}$, and consider a simple lemma:

Lemma 7 *In semimodules over entire, zerosumfree semirings, only vectors with empty intersection of supports are orthogonal.*

Proof Suppose $v \perp u$, then $\sum_{i=1}^{n} v_i \otimes u_i = \epsilon$. If any $v_i = \epsilon$ or $u_i = \epsilon$ then their product is null, so we need only consider a non-empty $\mathrm{supp}(v) \cap \mathrm{supp}(u)$. In this case, $v^{\mathsf{T}} \otimes u = \sum_{i \in \mathrm{supp}(v) \cap \mathrm{supp}(u)} v_i \otimes u_i$. But if $\mathscr{S}$ is zerosumfree, for the sum to be null every factor has to be null. And for a factor to be null, since $\mathscr{S}$ is entire, either v_i is null, or u_i is null, and then i would not belong to the common support. □

2.4.2 The Null Eigenspaces

If any, the eigenvectors of the null eigenvalue are interesting in that they define the null eigenspace. Also, the particular eigenvalue $\perp$ can only appear in UFNF$_3$. The following proposition describes the null eigenvalue and eigenspace:

Proposition 1 *Let $\overline{\mathscr{S}}$ be a semiring and $A \in \mathscr{M}_n(\overline{\mathscr{S}})$. Then:*

1. *If the i-th column is zero then the characteristic vector e_i is a fundamental eigenvector of ϵ for A and $\epsilon \in \mathrm{P}^{\mathrm{p}}(A)$.*
2. *Non-collinear eigenvectors of ϵ are orthogonal, so the order of multiplicity of $\perp \in \mathrm{P}^{\mathrm{p}}(A)$ is the number of empty columns of A.*
3. *If $\mathscr{S}$ is entire, then G_A has no cycles if and only if $\mathrm{P}^{\mathrm{p}}(A) = \{\epsilon\}$.*
4. *If $\mathscr{S}$ is entire and zerosumfree, the null eigenspace if generated by the fundamental eigenvectors of ϵ for A, $\mathscr{V}_\epsilon(A) = \langle \mathrm{FEV}_\rho(\epsilon) A \rangle$.*

Proof See [35, 3.6 and 3.7]. Claim 2 is a consequence of claim 1 and Lemma 7. □
Note that these are important in as much as they generate $\perp$ coordinates in the eigenvectors, that is, in the hubs and authorities vectors.

2.4.3 Eigenvalues and Eigenvectors of Matrices over Positive Semifields

When $\mathcal{S}$ has more structure we can improve on the results in the previous section. The first proposition advises us to concentrate on the irreducible blocks:

Proposition 2 *If $\mathcal{K}$ is a positive semifield, $A \in \mathcal{M}_n(\mathcal{K})$ is irreducible, and $v \in \mathcal{V}_\rho(A)$ then $\rho > \epsilon$ and $\forall i \in \bar{n}$, $v_i > \epsilon$.*

Proof See [33, Lemma 4.1.2]. □

Note that these results apply to $\mathbb{R}_0^+$, but not to $\mathbb{R}_{+,\times}$, the reals, or to $\mathbb{C}_{+,\times}$, the complex numbers, since the latter are *not* dioids.

For a finite $\rho \in K$ in a semifield, let $(\widetilde{A^\rho})^+ = (\rho^{-1} \otimes A)^+$ be the *normalized transitive closure* of A. The lemma below allows us to change the focus from the transitive closures to the circuit structure of G_A and vice versa.

Lemma 8 *If $\mathcal{K}$ is a semifield and $A \in \mathcal{M}_n(\mathcal{K})$, then if $(\rho^{-1} \otimes A)^*$ exists and if either $\sum_{c \in C_i} w(c) \otimes (\rho^{-1})^{l(c)} \oplus e = \sum_{c \in C_i} w(c) \otimes (\rho^{-1})^{l(c)}$ where C_i denotes the set of circuits in C_A^+ containing node v_i, or $(\rho^{-1} \otimes A)_{\cdot i}^* = (\rho^{-1} \otimes A)_{\cdot i}^+$ then $(\rho^{-1} \otimes A)_{\cdot i}^*$ is an eigenvector of A for eigenvalue ρ.*

Proof See [33, Chapter 6, Corollary 2.4]. □

When $\mathcal{K}$ is a radicable semifield, the *mean of cycle* c is $\mu_\oplus(c) = \sqrt[l(c)]{w(c)}$, If the semifield is (additively) idempotent the *aggregate cycle mean* of A is $\mu_\oplus(A) = \sum\{\mu_\oplus(c) \mid c \in C_A^+\}$. If the semiring is idempotent and selective, the nodes in the circuits that attain this mean are called the *critical nodes of A*, $V_A^c = \{i \in c \mid \mu_\oplus(c) = \mu_\oplus(A)\}$. Then the critical nodes are $V_A^c = \{i \in V_A \mid (\widetilde{A^+})_{ii}^+ = e\}$.

We define the set of (right) *fundamental eigenvectors of A for ρ* as those indexed by the critical nodes.

$$\mathrm{FEV}_\rho(A) = \{(\widetilde{A})_{\cdot i}^+ \mid i \in V_A^c\} = \{(\widetilde{A})_{\cdot i}^+ \mid (\widetilde{A})_{ii}^+ = e\}.$$

The basic building block is the spectrum of irreducible matrices:

Theorem 4 ((Right) Spectral Theory for Irreducible Matrices [35]) *Let $A \in \mathcal{M}_n(\mathcal{K})$ be an irreducible matrix over a complete commutative selective radicable semifield. Then:*

1. *The right spectrum of the matrix includes the whole semiring but the zero:*

$$P(A) = \overline{\mathcal{K}} \setminus \{\bot\}$$

2. *The right proper spectrum only comprises the aggregate cycle mean:*

$$P^p(A) = \{\mu_\oplus(A)\}$$

3. *If an eigenvalue is improper $\rho \in P(A) \setminus P^p(A)$, then its eigenspace (and eigenlattice) is reduced to the two vectors:*

$$\mathcal{V}_\rho(A) = \{\perp^n, \top^n\} = \mathcal{L}_\rho(A)$$

4. *The eigenspace for a finite proper eigenvalue $\rho = \mu_\oplus(A) < \top$ is generated from its fundamental eigenvectors over the whole semifield, while the eigenlattice is generated by the semifield $\mathbb{K} = \langle \{\perp, e, \top\}, \oplus, \otimes, \perp, e, \top\rangle$.*

$$\mathcal{V}_\rho(A) = \langle \mathrm{FEV}_\rho(A)\rangle_{\overline{\mathcal{K}}} \supset \mathcal{L}_\rho(A) = \langle \mathrm{FEV}_\rho(A)\rangle_{\mathbb{K}}$$

Note how this theorem introduces the notion of eigenlattices to finitely represent an eigenspace over an idempotent semifield. Refer to [35] for further details.

We will see in our results that the only other UFNF type we need be concerned about is UFNF$_2$: Let the partition of V_A generating the permutation that renders A in UFNF$_2$, block diagonal form, be $V_A = \{V_k\}_{k=1}^K$, and write $A = \biguplus_{k=1}^K A_k$, $A_k = A(V_k, V_k)$.

Lemma 9 *Let $A = \biguplus_{k=1}^K A_k \in \mathcal{M}_n(\mathcal{S})$ be a matrix in UFNF$_2$, over a semiring, and $\mathcal{V}_\rho(A_k)$ $(\mathcal{U}_\lambda(A_k))$ a right (left) eigenspace of A_k for ρ (λ). Then,*

$$\mathcal{U}_\lambda(A) \cong \underset{k=1}{\overset{K}{\times}} \mathcal{U}_\lambda(A_k) \qquad\qquad \mathcal{V}_\rho(A) \cong \underset{k=1}{\overset{K}{\times}} \mathcal{V}_\rho(A_k). \qquad (15)$$

Proof See [36] Lemma 3.12.   □

Note that this procedure is constructive and how the combinatorial nature of the proof in [36] makes the claim hold in any semiring. Clearly, if $\rho \in \mathrm{P}^p(A_k)$ for any k, then $\rho \in \mathrm{P}^p(A)$. Since $\mathrm{P}^p(A_k) = \Lambda^p(A_k)$ for matrices admitting an UFNF$_2$, $\mathrm{P}^p(A_k) = \Lambda^p(A_k) = \bigcup_{k=1}^K \mathrm{P}^p(A_k)$.

Corollary 1 *Let $A \in \mathcal{M}_n(\mathcal{S})$ be a matrix in UFNF$_2$ over a semiring. Then the (left) right eigenspace of A for $\rho \in \mathrm{P}(A)$ is the product of the (left) right eigenspaces for the blocks, $\mathcal{U}_\lambda(A) = \times_{k=1}^K \mathcal{U}_\lambda(A_k)$ and $\mathcal{V}_\rho(A) = \times_{k=1}^K \mathcal{V}_\rho(A_k)$.*

In complete semirings, looking for generators for the eigenspaces with $\delta_k(k) = e$ and $\delta_k(i) = \perp$ for $k \neq i$, we define the right fundamental eigenvectors as

$$\mathrm{FEV}_\rho^2(A) = \bigcup_{k=1}^K \left[\underset{i=1}{\overset{K}{\times}} \delta_k(i) \otimes \mathrm{FEV}_\rho^1(A_i) \right]. \qquad (16)$$

Lemma 9 proves that $\mathrm{FEV}_\rho^2(A) \subset \mathcal{V}_\rho(A)$, but we also have the following:

Lemma 10 *Let $A \in \mathcal{M}_n(\mathcal{D})$ be a matrix in UFNF$_2$ over a complete idempotent semiring with $\rho \in \mathrm{P}(A)$. Then,*

1. *If $\rho \in \mathrm{P}^p(A)$, then $\mathrm{FEV}_\rho^{2,\mathrm{F}}(A) = \bigcup_{k|\rho \in \mathrm{P}^p(A_k)} \left[\times_{i=1}^K \delta_k(i) \otimes \mathrm{FEV}_\rho^{1,\mathrm{F}}(A_i) \right]$.*
2. *If $\rho \in \mathrm{P}(A) \setminus \mathrm{P}^p(A)$ then $\mathrm{FEV}_\rho^2(A) = \mathrm{FEV}^{2,\top}(A)$.*
3. *If $\rho \in \mathrm{P}^p(A)$ then $\mathrm{FEV}_\rho(A) = \mathrm{FEV}_\rho^{2,\mathrm{F}}(A) \cup \mathrm{FEV}^{2,\top}(A) \setminus \top \otimes \mathrm{FEV}_\rho^{2,\mathrm{F}}(A)$.*

4. $\text{FEV}^{2,\top}_{\cdot}(A) = \top \otimes \text{FEV}^2_\rho(A).$

So call $\text{FEV}^{2,\top}(A)$ the *saturated fundamental eigenvectors of A*, and define the *(right) saturated eigenspace* as $\mathscr{V}^\top(A) = \langle \text{FEV}^{2,\top}(A) \rangle_{\overline{\mathscr{D}}}.$

Corollary 2 *Let $A \in \mathcal{M}_n(\mathscr{S})$ be a matrix in UFNF_2 over a complete selective radicable idempotent semifield. Then*

1. *For $\rho \in P^P(A)$, $\mathscr{V}_\rho(A) \supseteq \mathscr{V}^\top(A)$.*
2. *For $\rho \in P(A) \setminus P^P(A)$, $\mathscr{V}_\rho(A) = \mathscr{V}^\top(A)$.*

Notice that the very general proposition below is for *all* complete dioids.

Proposition 3 *Let $A \in \mathcal{M}_n(\overline{\mathscr{D}})$ be a matrix in UFNF_2 over a complete dioid. Then,*

1. *For $\rho \in P(A) \setminus P^P(A)$,*

$$\mathscr{U}^\top(A) = \langle \text{FEV}^{2,\top}(A^\top) \rangle_{\mathrm{I\!K}} \qquad \mathscr{V}^\top(A) = \langle \text{FEV}^{2,\top}(A) \rangle_{\mathrm{I\!K}}.$$

2. *For $\rho \in P^P(A)$, $\rho < \top$,*

$$\mathscr{U}_\lambda(A) = \langle \text{FEV}^2_\rho(A^\top) \rangle_{\overline{\mathscr{D}}} \qquad \mathscr{V}_\rho(A) = \langle \text{FEV}^2_\rho(A) \rangle_{\overline{\mathscr{D}}}.$$

To better represent eigenspaces, we define the *spectral lattices of A*,

$$\mathscr{L}_\lambda(A) = \langle \text{FEV}^2_\rho(A^\top)^\top \rangle_{\mathrm{I\!K}} \qquad \mathscr{L}_\rho(A) = \langle \text{FEV}^2_\rho(A) \rangle_{\mathrm{I\!K}}.$$

involving the product of the component lattices, $\mathscr{L}_\rho(A) = \bigtimes_{k=1}^{K} \mathscr{L}_\rho(A_k).$

3   Results

3.1   HITS over Idempotent Semifields: iHITS

Let $\overline{\mathscr{K}} = \langle K, \oplus, \otimes, \bot \rangle$ be a complete dioid, in general, and let (G, M, R) be a $\mathscr{K}$-formal context. Then the space of hub scores is $\mathscr{X} = K^G$ and that of authorities is $\mathscr{Y} = K^M$ and they get mutually transformed by the actions of two linear functions:

$$R^\top \otimes \cdot : K^G \to K^M \qquad\qquad R \otimes \cdot : K^M \to K^G$$

$$x \mapsto R^\top \otimes x \qquad\qquad y \mapsto R \otimes y \qquad (17)$$

To relate this problem back to the original one, we rewrite (7) in the new spaces,

$$A \otimes z = z \otimes \sigma \Leftrightarrow \begin{bmatrix} \cdot & R \\ R^{\mathsf{T}} & \cdot \end{bmatrix} \otimes \begin{bmatrix} x \\ y \end{bmatrix} = \begin{bmatrix} x \\ y \end{bmatrix} \otimes \sigma \tag{18}$$

we premultiply (18) by the new symmetric A to obtain

$$(R \otimes R^{\mathsf{T}}) \otimes h = h \otimes \sigma^{\otimes 2} \qquad\qquad (R^{\mathsf{T}} \otimes R) \otimes a = a \otimes \sigma^{\otimes 2} \tag{19}$$

expressing the solution in terms of the projectors on both modes of (1). This proves that in order to solve the HITS problem in a dioid we have to solve the Singular Value Problem (18) which amounts to solving both decoupled Eigenvalue Problems (19).

However, these manipulations have overlooked the fact that to obtain the original HITS an idempotent semifield—not a dioid—was needed. In the interest of generality, we will develop the calculations below in terms of dioids, but the reader is warned that after a certain point we must suppose that $\mathcal{K}$ is a semifield to reach a solution, and in particular an idempotent semifield. In this way we obtain the analog of HITS over idempotent semifields which we call iHITS.

3.2   The Eigenproblem in Symmetric Matrices

Since A is symmetric, we no longer have to worry about the distinction between the left and right eigenproblem:

Lemma 11 *If A is symmetric then $\Lambda^{\mathrm{P}}(A_k) = \mathrm{P}^{\mathrm{P}}(A_k)$ and $\left(\mathcal{U}_\rho(A)\right)^{\mathsf{T}} = \mathcal{V}_\rho(A)$.*

We can also refine the results in Proposition 1:

Proposition 4 *Let $\overline{\mathscr{S}}$ be a semiring and a symmetric $A \in \mathcal{M}_n(\overline{\mathscr{S}})$. Then:*

1. *The multiplicity of $\perp \in \mathrm{P}^{\mathrm{P}}(A)$ is the number of empty rows/columns of A.*
2. *If $\mathscr{S}$ is entire, $\mathrm{P}^{\mathrm{P}}(A) = \{\epsilon\}$ if and only if $A = \mathcal{E}_n$.*

Proof First, if $A = A^{\mathsf{T}}$, then the number of empty rows and empty columns is the same, and Proposition 1.2 provides the result. Second, after Proposition 1.3 $\mathrm{P}^{\mathrm{P}}(A) = \{\epsilon\}$ means G_A has no cycles. But if $A_{ij} \neq \epsilon$ then $c = i \to j \to i$ is a cycle with non-null weight $w(c) = A_{ij} \otimes A_{ji} \neq \epsilon$, which is a contradiction. $\square$

Note that empty rows of R generate left eigenvalues while empty columns generate right eigenvalues, so the multiplicity of the null singular value may change from left to right.

To use Lemma 8 and Theorem 4, we need the maximum cycle mean:

Proposition 5 *Let $\mathcal{K}$ be a complete idempotent semifield and let $A \in \mathcal{M}_n(K)$ be symmetric. Then $\mu_\oplus(c) = \sup_{i,j} A_{ij}$, where the sup, taken in the natural order of the semifield is attained.*

Proof Since A is symmetric, $c = i \to j \to i$ is a cycle whenever $A_{ij} = A_{ji} \neq \perp$. Then $\mu_\oplus(c) = A_{ij}$. Consider one c' such that $\mu_\oplus(c') = \sup_{i,j} A_{ij} = \max_{i,j} A_{i,j}$ in the order of the semiring. This must exist since i, j are finite. If we can extend any of these critical cycles with another node k such that $c'' = i \to j \to k \to i$ then $w(c'') = A_{ij} \otimes A_{jk} \otimes A_{ki} \leq A_{ij}^3 = \mu_\oplus(c')^{l(c'')}$, so in the aggregate mean $\mu_\oplus(c') \oplus \mu_\oplus(c'') = \mu_\oplus(c')$. So we induce on the length of any cycle that is an extension of c' that $\sum_{c \in C_A^+} = \mu_\oplus(c') = \sup_{i,j} A_{i,j}$. □

To find the cycle means easily, we use the UFNF form.

3.3 UFNF Forms of Symmetric Matrices and Their Closures

For symmetric reducible matrices, the feasible UFNF types are simplified:

Proposition 6 *Let $\mathscr{S}$ be a dioid, and a symmetric $A \in \mathscr{M}_n(\mathscr{S})$. Then:*

- *(UFNF$_3$) A admits a proper symmetric UFNF$_3$ form if it has zero lines, and, in that case, the set of zero lines and zero rows are the same.*

$$P_3^T \otimes A \otimes P_3 = \begin{bmatrix} A_{\beta\beta} & \cdot \\ \cdot & \mathscr{E}_{\iota\iota} \end{bmatrix} \quad (20)$$

- *(UFNF$_2$) If a A has no zero lines it can be transformed by a simultaneous row and column permutation $P_2 = P(A_1) \ldots P(A_k)$ into symmetric block diagonal UFNF:*

$$P_2^T \otimes A \otimes P_2 = \begin{bmatrix} A_1 & \cdot & \ldots & \cdot \\ \cdot & A_2 & \ldots & \cdot \\ \vdots & \vdots & \ddots & \vdots \\ \cdot & \cdot & \ldots & A_K \end{bmatrix} = \biguplus_{k=1}^{K} A_k \quad (21)$$

where $\{A_k\}_{k=1}^K$, $K \geq 1$ are the symmetric matrices of connected components of $\overline{G}_A$.
- *(no UFNF$_1$) A cannot be permuted into a proper UFNF$_1$ form.*

Proof A simple matrix conformation procedure on Lemma 6 when the matrix is symmetric. □

We will see that this is almost the only structure we need to consider to find the eigenvectors. Consider $R \in \mathscr{M}_{G+M}(S)$, the bipartite network matrix, for instance, of the form:

$$R = \begin{bmatrix} R_1 & R_{12} \\ R_{21} & R_2 \end{bmatrix} \quad (22)$$

If R_{12} and R_{21} are null, then we can find a permutation P so that

$$P^{\mathrm{T}} \otimes A \otimes P = P^{\mathrm{T}} \otimes \begin{bmatrix} \cdot & \cdot & R_1 & \cdot \\ \cdot & \cdot & \cdot & R_2 \\ R_1^{\mathrm{T}} & \cdot & \cdot & \cdot \\ \cdot & R_2^{\mathrm{T}} & \cdot & \cdot \end{bmatrix} \otimes P = \begin{bmatrix} \cdot & R_1 & \cdot & \cdot \\ R_1^{\mathrm{T}} & \cdot & \cdot & \cdot \\ \cdot & \cdot & \cdot & R_2 \\ \cdot & \cdot & R_2^{\mathrm{T}} & \cdot \end{bmatrix} \tag{23}$$

Now, if $R_2 = \mathscr{E}_{G_2 M_2}$ is null, then (23) is in UFNF$_3$ with $\mathscr{E}_u = \mathscr{E}_{(G_2+M_2)(G_2+M_2)}$, while if R_1 and R_2 are both full, then (23) is in UFN$_2$ with blocks A_1 and A_2, respectively. Note that other blocked forms of R simply generate an irreducible A, since the UFNF$_1$ form is not possible.

So we can suppose that A can be simultaneously row and column permuted into a diagonal block form

$$P^{\mathrm{T}} \otimes A \otimes P = \begin{bmatrix} A_1 & \cdots & \cdot & & \cdot \\ \vdots & \ddots & \vdots & & \vdots \\ \cdot & \cdots & A_K & & \cdot \\ \cdot & \cdots & \cdot & & \mathscr{E}_u \end{bmatrix} = \left(\biguplus_{k=1}^{K} A_k \right) \uplus \mathscr{E}_u \quad A_k = \begin{bmatrix} \cdot & R_k \\ R_k^{\mathrm{T}} & \cdot \end{bmatrix} \tag{24}$$

with the empty lines and rows permuted to the beginning $\mathscr{E}_u$ and irreducible blocks A_k. Recall that closures and permutations commute, whence the closures of the matrices in the forms above are really simple: the closure of (24) is straightforward in terms of the closures of the blocks:

$$P^{\mathrm{T}} \otimes A^+ \otimes P = \begin{bmatrix} A_1^+ & \cdots & \cdot & & \cdot \\ \vdots & \ddots & \vdots & & \vdots \\ \cdot & \cdots & A_K^+ & & \cdot \\ \cdot & \cdots & \cdot & & \mathscr{E}_u \end{bmatrix} \tag{25}$$

The solution of this base case is highly dependent in the dioid in which the problem is stated. Since we will be solving the problem in idempotent semifields, for the irreducible base case we need only be concerned about matrix:

$$\widetilde{A}^{\mu \oplus (A)} = \begin{bmatrix} \cdot & \widetilde{R}^{\mu \oplus (A)} \\ \left(\widetilde{R}^{\mu \oplus (A)} \right)^{\mathrm{T}} & \cdot \end{bmatrix} = \begin{bmatrix} \cdot & B \\ B^{\mathrm{T}} & \cdot \end{bmatrix} \tag{26}$$

where we are using the shorthand $B = \widetilde{R}^{\mu \oplus (A)}$ to account for the normalization with the cycle means that we need to use to find the eigenvectors. To find the closures of such irreducible matrices apply Lemma 3 to (26):

$$\left(\widetilde{A}^{\mu \oplus (A)} \right)^+ = \begin{bmatrix} (B \otimes B^{\mathrm{T}})^+ & B \otimes (B^{\mathrm{T}} \otimes B)^* \\ B^{\mathrm{T}} \otimes (B \otimes B^{\mathrm{T}})^* & (B^{\mathrm{T}} \otimes B)^+ \end{bmatrix} \tag{27}$$

where we have used that $(B^{\mathrm{T}} \otimes B)^* \otimes B^{\mathrm{T}} = B^{\mathrm{T}} \otimes (B \otimes B^{\mathrm{T}})^*$.

3.4 Pairs of Singular Vectors of Symmetric Matrices over Idempotent Semifields

To extract the eigenvectors corresponding to left singular vectors $i \in G$ (respectively, right singular vectors $j \in M$) we need to check Theorem 3 against each block in its diagonal:

$$i \in G, \left(\widetilde{A}^{\mu \oplus (A)}\right)^*_{\cdot i} = \left(\widetilde{A}^{\mu \oplus (A)}\right)^+_{\cdot i} \Leftrightarrow (B \otimes B^{\mathrm{T}})^*_{\cdot i} = (B \otimes B^{\mathrm{T}})^+_{\cdot i}$$

$$j \in M, \left(\widetilde{A}^{\mu \oplus (A)}\right)^*_{\cdot j} = \left(\widetilde{A}^{\mu \oplus (A)}\right)^+_{\cdot j} \Leftrightarrow (B^{\mathrm{T}} \otimes B)^*_{\cdot j} = (B^{\mathrm{T}} \otimes B)^+_{\cdot j}$$

So the existence of $\left(\widetilde{A}^{\mu \oplus (A)}\right)^+$ requires the existence of the transitive closures $(B^{\mathrm{T}} \otimes B)^+$ and $(B \otimes B^{\mathrm{T}})^+$ as we could have expected from (19). Note that after (6), the hub and authority scores are those columns such that:

$$(B \otimes B^{\mathrm{T}})^+_{\cdot i} = (B \otimes B^{\mathrm{T}})^*_{\cdot i} \qquad\qquad (B^{\mathrm{T}} \otimes B)^+_{\cdot j} = (B^{\mathrm{T}} \otimes B)^*_{\cdot j}$$

but, importantly, (27) gives us the form of the authority score related to a particular hub score and vice versa, which is a kind of formal-concept property:

$$h_i = (B \otimes B^{\mathrm{T}})^*_{\cdot i} \Leftrightarrow a_i = B^{\mathrm{T}} \otimes (B \otimes B^{\mathrm{T}})^*_{\cdot i} \tag{28}$$

$$a_j = (B^{\mathrm{T}} \otimes B)^*_{\cdot j} \Leftrightarrow h_j = B \otimes (B^{\mathrm{T}} \otimes B)^*_{\cdot j}$$

This solves completely the description of the left and right singular vectors. To find the singular values, we note from (19) that they are the square roots of the cycle means of the independent blocks or, equally, the proper eigenvalues of A,

$$\Sigma = \{ \sqrt{\rho} \mid \rho = \mu_\oplus(A_k), A = \oplus_k A_k \} = \{ \sqrt{\rho} \mid \rho = \mathrm{P}^{\mathrm{P}}(A_k) \}$$

This would include the bottom if and only if one of the blocks is empty.

3.5 Relationship to FCA

In order to interpret the results above in the light of FCA, we have to use the proper multi-valued extension of it. For such purpose, $\mathscr{K}$-Formal Concept Analysis is an extension of FCA for formal contexts with entries in an idempotent semifield [24] which has been used for the analysis of confusion matrices and other data with the appropriate characteristics [39–41].

Regarding the glimpse of a formal concept-like property of (28), a cursory analysis with the techniques in [24] shows that the analogue of the polars in FCA declared in (17) actually comes from two different generalized Galois connections:

- $R \otimes \cdot : K^M \to K^G$ is the right adjunct of a left (Galois) adjunction, while
- $R^T \otimes \cdot : K^G \to K^M$ is the left adjunct in a right (Galois) adjunction.

It is well known that the iteration of these operators in the Boolean case is not concept-forming. Nevertheless, they do lead to closures. For instance, in [15] a discussion of the issue leads to operators achieving transitive closures for endorelations—that is, binary relations with identical domain and codomain—and the finding of strongly connected components in the one-mode projection graphs these endorelations define. This agrees with a technical condition often imposed on graphs prior to their study through HITS: that their adjacency matrices be irreducible, which translates into a graph with a single strongly connected component. Note also that the transitive closure of the matrix also figures prominently in this application.

Yet, in this chapter the isomorphism between the ranges of the operators in (17) defining the duality between hubs and authorities is clear and attested in a more generic context than HITS was initially conceived. Specifically, we consider graphs with any number of strongly connected components, even with none (see Sect. 3.3).

In parallel work, also, we suggest that the "standard" take on what a formal concept of a $\mathscr{K}$-context is should be enlarged to include not only closures, but also the interior of (multi-valued) sets of objects and attributes [42]. It may be the case that the hub and authority score vectors in the idempotent version of HITS belong to these systems of interiors. In any case, it would seem that there is not a single SVD for matrices with values in an idempotent semifield, and this issue needs to be explored further.

3.6 Example

In this section we present a HITS analysis for a weighted two-mode network both using standard HITS and HITS over the max-min-plus idempotent semifield. The data being analysed is the example in [43, p. 31], the worries data, which is a two-mode network of the type of worry declared as most prevalent by 1554 adult Israeli depending on their living countries—and sometimes those of their parents. The graph of the network is depicted in Fig. 1a.

To be amenable for max-min-plus processing the original counts in the contingency matrix were transformed into a joint probability function P_{WP} with marginals P_W over worries and P_P over procedence. The idempotent SVD was carried out on the pointwise mutual information matrix

$$I_{WP}(i,j) = \frac{P_{WP}(i,j)}{P_W(i) \cdot P_P(j)} \; .$$

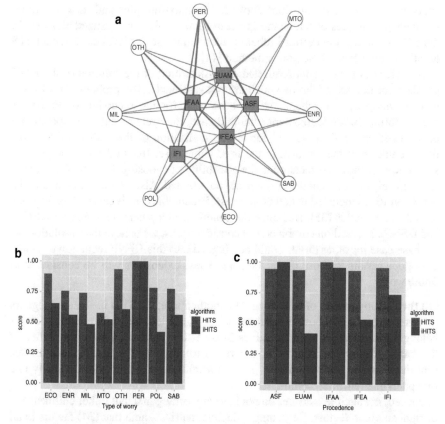

Fig. 1 Weighted, directed graph of the worries data [43, p. 31] and its weighted idempotent and standard "authority" (worry) and "hub" (procedence) scores. There are clear differences in both approaches. (**a**) Worries weighted bipartite network. (**b**) Principal worry scores. (**c**) Procedence scores

The "authority" and "hub" scores are differentiated for each of the modes: they return "type of worry" and "procedence" scores, respectively. We can see that HITS and iHITS produce somehow different results: the actual meaning of these differences is data dependent and a matter for future, more specialized analyses. The idempotent primitives were developed in-house and are available from the authors upon request.

3.7 Discussion

Extensions to Other Semirings Note that the original HITS problem was set in a positive semifield that is not idempotent; hence, our method of solution does not apply yet to that case, but does apply to the max-min-plus and max-min-times semifields, examples of which are given in terms of their normalized closures. In the case of the $\mathbb{R}_0^+$ the Perron-Frobenius theorem is usually invoked to solve HITS iteratively by means of the power method [44].

A reviewer of this paper requested a consideration of the solution of the HITS problem for the rest of dioids which are not semifields. The problem with further generalization of our scheme is the base case for the recursion of Frobenius normal forms. In the idempotent semifield case, the cycle means of Proposition 5 provide the eigenvalues needed for the normalization of the matrices that allow the calculation of the closures in the irreducible case. But in the generic positive semifield case the cycle means and the possibility of choosing the critical circuits to select the eigenvectors in the closures are not granted. Also, in inclines—and in the fuzzy semirings included in them—the base, irreducible case is completely different to that of semifields [33]. But since the generic development on dioids for UFNF1 and UFNF2 is based on combinatorial considerations, we believe that a solution for the base case for other dioids could be plugged into this UFNF recursion to obtain analogous results to those presented here. These extensions will be considered in future work.

On the Orthogonality of Solutions On another note, the SVD in standard algebra makes a strong case about the orthogonality of the left and right singular vectors belonging to different singular values in order to guarantee certain properties of the bases of singular vectors in the reconstruction. But in entire zerosumfree semirings, and in entire dioids or positive semifields a fortiori, orthogonality is a rare phenomenon, after Lemma 7.

Indeed, irreducible matrices do not have any orthogonal, but rather collinear left or right singular vectors. Regarding reducible matrices, note that (24) factors in all the possible orthogonality between eigenvectors. In fact we have,

Corollary 3 *Let* (G, M, R) *be an* $\mathscr{S}$*-formal context over a dioid* $\mathscr{S}$*, and* $A \in \mathscr{M}_n(\mathscr{S})$ *as in* (27)*. Then two of the eigenvectors for A or (left, right) singular vectors for R can only be orthogonal if they arise from different blocks.*
and Proposition 3 proves that even in that case they might not be orthogonal.

However, after the work in [22, 25] orthogonality may not be needed in the case of dioids: the use of join- and meet-irreducible may guarantee perfect reconstruction.

On the Effectiveness of the Dual-Projection Approach for 2-Mode Network Analysis Yet another reviewer raised the concern that the work in [3] proves the dual-projection approach hopeless. This work of Latapy and colleagues propounds a more "direct" approach to the study of bipartite networks by means of collecting and creating measures designed specifically for them, as opposed to those adapted from 1-mode networks. They develop to some extent a criticism of the projection approach and, indirectly, of the dual-projection approach on *Boolean networks*.

However, they suggest that theirs and the projection approach are, in general, *complementary*. And in particular, that none of the criticism for Boolean projection approaches applies to projection approaches on weighted affiliation networks. On these grounds and the present work, the criticism of [3] notwithstanding, we believe the dual-projection approach is adequate for the study of 2-mode networks and can bring many insights about their behaviour.

4 Summary and Conclusions

In this paper, we related the HITS algorithms to the SVD of the adjacency matrix of a weighted 2-mode network and argued that this supports the dual-projection approach to SNA.

To make evident the relationship of these techniques to $\mathcal{K}$-Formal Concept Analysis, we generalized the HITS algorithms for semirings, then instantiated it in dioids, semifields (including the original semifield where it was defined) and finally in idempotent semifields, which are the algebras used by $\mathcal{K}$-FCA.

We showed that the projection operators are related to Galois adjunctions, rather than to the polars in Galois connections, and that this approach to weighted graph analysis has affinities to finding strongly connected components in Boolean graphs. What the connected components of weighted graphs might mean is subject for further work.

We have also provided an example of how to use the new calculations to obtain idempotent authority and hubness scores for a weighted bipartite graph, although the interpretation of such scores vis-à-vis the original ones needs further investigation.

Acknowledgements The authors have been partially supported by the Spanish Government-MinECo projects TEC2014-53390-P and TEC2014-61729-EXP for this work.

We would like to thank the reviewers of earlier versions of this paper for their help in improving it.

References

1. Bang-Jensen, J., Gutin, G.: Digraphs. Theory, Algorithms, and Applications, 3rd edn. Springer, Heidelberg (2001)
2. Agneessens, F., Everett, M.G.: Introduction to the special issue on advances in two-mode social networks. Soc. Netw. **35**, 145–147 (2013)
3. Latapy, M., Magnien, C., Vecchio, N.D.: Basic notions for the analysis of large two-mode networks. Soc. Netw. **30**, 31–48 (2008)
4. Everett, M.G., Borgatti, S.P.: The dual-projection approach for two-mode networks. Soc. Netw. **35**, 204–210 (2013)
5. Strang, G.: The fundamental theorem of linear algebra. Am. Math. Mon. **100**, 848–855 (1993)
6. Golub, G.H., Van Loan, C.F.: Matrix Computations, 3rd edn. JHU Press, Baltimore (2012)

7. Landauer, T.K., McNamara, D.S., Dennis, S., Kintsch, W.: Handbook of Latent Semantic Analysis. Lawrence Erlbaum Associates, Mahwah (2007)
8. Ganter, B., Wille, R.: Formal Concept Analysis: Mathematical Foundations. Springer, Berlin/Heidelberg (1999)
9. Wille, R.: Restructuring lattice theory: an approach based on hierarchies of concepts. In: Ordered Sets (Banff, Alta., 1981), pp. 445–470. Reidel, Boston (1982)
10. Davey, B., Priestley, H.: Introduction to Lattices and Order, 2nd edn. Cambridge University Press, Cambridge (2002)
11. Birkhoff, G.: Lattice Theory, 3rd edn. American Mathematical Society, Providence (1967)
12. Rota, G.C.: Indiscrete Thoughts. Springer, Boston, MA (2009)
13. Freeman, L.C., White, D.R.: Using Galois lattices to represent network data. Sociol. Methodol. **23**, 127–146 (1993)
14. Domenach, F.: CryptoLat - a pedagogical software on lattice cryptomorphisms and lattice properties. In: Ojeda-Aciego, M., Outrata, J. (eds.) 10th International Conference on Concept Lattices and Their Applications (2013)
15. Gaume, B., Navarro, E., Prade, H.: A parallel between extended formal concept analysis and bipartite graphs analysis. In: IPMU'10: Proceedings of the Computational Intelligence for Knowledge-Based Systems Design, and 13th International Conference on Information Processing and Management of Uncertainty, Universite Paul Sabatier Toulouse III. Springer, Heidelberg (2010)
16. Kuznetsov, S.O.: Interpretation on graphs and complexity characteristics of a search for specific patterns. Nauchno-Tekhnicheskaya Informatsiya Seriya - Informationnye i sistemy **1**, 23–27 (1989)
17. Falzon, L.: Determining groups from the clique structure in large social networks. Soc. Netw. **22**, 159–172 (2000)
18. Ali, S.S., Bentayeb, F., Missaoui, R., Boussaid, O.: An efficient method for community detection based on formal concept analysis. In: Foundations of Intelligent Systems, pp. 61–72. Springer, New York (2014)
19. Roth, C., Bourgine, P.: Epistemic communities: description and hierarchic categorization. Math. Popul. Stud. **12** 107–130 (2005)
20. Freeman, L.C.: Methods of social network visualization. In: Meyers, R.A. (ed.) Encyclopedia of Complexity and Systems Science, Entry 25, pp. 1–19. Springer, New York (2008)
21. Duquenne, V.: On lattice approximations: syntactic aspects. Soc. Netw. **18**, 189–199 (1996)
22. Bělohlávek, R., Vychodil, V.: Formal concepts as optimal factors in Boolean factor analysis: implications and experiments. In: Proceedings of the 5th International Conference on Concept Lattices and Their Applications, (CLA07), Montpellier, 24–26 October 2007
23. Bělohlávek, R.: Fuzzy Relational Systems. Foundations and Principles. IFSR International Series on Systems Science and Engineering, vol. 20. Kluwer Academic, Norwell (2002)
24. Valverde-Albacete, F.J., Peláez-Moreno, C.: Extending conceptualisation modes for generalised Formal Concept Analysis. Inf. Sci. **181**, 1888–1909 (2011)
25. Bělohlávek, R.: Optimal decompositions of matrices with entries from residuated lattices. J. Log. Comput. **22** (2012) 1405–1425
26. Kleinberg, J.M.: Authoritative sources in a hyperlinked environment. J. ACM **46** (1999) 604–632
27. Easley, D.A., Kleinberg, J.M.: Networks, Crowds, and Markets - Reasoning About a Highly Connected World. Cambridge University Press, Cambridge (2010)
28. Kolaczyk, E.D., Csárdi, G.: Statistical Analysis of Network Data with R. Use R!, vol. 65. Springer, New York, NY (2014)
29. Lanczos, C.: Linear Differential Operators. Dover, New York (1997)
30. Golan, J.S.: Power Algebras over Semirings. With Applications in Mathematics and Computer Science. Mathematics and Its applications, vol. 488. Kluwer Academic, Dordrecht (1999)

31. Pouly, M., Kohlas, J.: Generic Inference. A Unifying Theory for Automated Reasoning. Wiley, Hoboken (2012)
32. Golan, J.S.: Semirings and Their Applications. Kluwer Academic, Dordrecht (1999)
33. Gondran, M., Minoux, M.: Graphs, Dioids and Semirings. New Models and Algorithms. Operations Research/Computer Science Interfaces. Springer, New York (2008)
34. Gondran, M., Minoux, M.: Valeurs propres et vecteurs propres dans les dioïdes et leur interprétation en théorie des graphes. EDF, Bulletin de la Direction des Etudes et Recherches, Serie C, Mathématiques Informatique **2**, 25–41 (1977)
35. Valverde-Albacete, F.J., Peláez-Moreno, C.: The spectra of irreducible matrices over completed idempotent semifields. Fuzzy Sets Syst. **271**, 46–69 (2015)
36. Valverde-Albacete, F.J., Peláez-Moreno, C.: The spectra of reducible matrices over complete commutative idempotent semifields and their spectral lattices. Int. J. Gen. Syst. **45**, 86–115 (2016)
37. Valverde-Albacete, F.J., Peláez-Moreno, C.: Spectral lattices of reducible matrices over completed idempotent semifields. In: Ojeda-Aciego, M., Outrata, J., (eds.) Concept Lattices and Applications (CLA 2013), pp. 211–224. Université de la Rochelle, Laboratory L31, La Rochelle (2013)
38. Cohen, G., Gaubert, S., Quadrat, J.P.: Duality and separation theorems in idempotent semimodules. Linear Algebra Appl. **379**, 395–422 (2004)
39. Peláez-Moreno, C., García-Moral, A.I., Valverde-Albacete, F.J.: Analyzing phonetic confusions using Formal Concept Analysis. J. Acoust. Soc. Am. **128**, 1377–1390 (2010)
40. Valverde-Albacete, F.J., González-Calabozo, J.M., Peñas, A., Peláez-Moreno, C.: Supporting scientific knowledge discovery with extended, generalized formal concept analysis. Expert Syst. Appl. **44**, 198–216 (2016)
41. González-Calabozo, J.M., Valverde-Albacete, F.J., Peláez-Moreno, C.: Interactive knowledge discovery and data mining on genomic expression data with numeric formal concept analysis. BMC Bioinf. **17**, 374 (2016)
42. Valverde-Albacete, F.J., Peláez-Moreno, C.: The linear algebra in extended formal concept analysis over idempotent semifields. In: Bertet, K., Borchmann, D. (eds.) Formal Concept Analysis, Springer Berlin Heidelberg, 211–227 (2017)
43. Mirkin, B.: Mathematical Classification and Clustering. Nonconvex Optimization and Its Applications, vol. 11. Kluwer Academic, Dordrecht (1996)
44. Akian, M., Gaubert, S., Ninove, L.: Multiple equilibria of nonhomogeneous Markov chains and self-validating web rankings. arXiv:0712.0469 (2007)

Printed in the United States
By Bookmasters